GEOMETRICAL
Quickies & Trickies

Over 200 Trick & Tricky Geometry
to Enhance Your Problem-Solving Skills

K. C. Yan

MATHPLUS Publishing
Blk 639 Woodlands Ring Road
#02-35 Singapore 730639

E-mail: publisher@mathpluspublishing.com
Website: www.mathpluspublishing.com

MathPlus Publishing books are available at special discounts for bulk purchases for mathematics conferences or school use. For more information, contact the Sales and Marketing Department at (65) 63646077 or email publisher@mathpluspublishing.com.

National Library Board, Singapore Cataloguing in Publication Data

Names: Yan, Kow Cheong, author.
Title: Geometrical quickies & trickies / K. C. Yan.
Description: Singapore : MathPlus Publishing, 2016. | "Over 200 trick & tricky geometry to enhance your problem solving skills."
Identifiers: OCN 930454451 | ISBN 978-981-09-7389-6 | (paperback)
Subjects: LCSH: Geometry--Problems, exercises, etc. | Geometry--Miscellanea. | Mathematical recreations.
Classification: LCC QA459 | DDC 516.0076--dc23

Printed in the United States of America

Preface

When *Mathematical Quickies & Trickies* was released in 1998, it somewhat stood out from the crowd of drill-and-kill assessment mathematics titles. And more than one and a half decades later, this problem solving book continues to serve well those who are looking for some creative nonroutine questions to challenge themselves mathematically.

In the spirit of *Mathematical Quickies & Trickies*, which focuses more on arithmetic and a bit on algebra, **Geometrical Quickies & Trickies** aims to meet the geometrical needs and wants of problem solvers and mathletes, preparing for mathematics contests and competitions. The 200+ trick and tricky geometry questions will help those, bored by sterile questions found in most local textbooks and assessment titles, to stretch and enhance their problem-solving and visualization skills.

For those new to the terms "mathematical quickies" and "mathematical trickies," allow me to redefine them here:

- A *mathematical quickie* is a problem which may be solved by tedious methods, but with which with proper insight may be disposed of quickly.

- A *mathematical trickie* is a problem whose solution lies on some key word, phrase, or idea rather than on a mathematical routine.

So, a *geometrical quickie* may be defined as a geometry problem, whose solution conventionally involves a tedious process, but which with proper insight may be solved relatively quickly.

Geometrical Quickies & Trickies has a wide variety of questions which would challenge both the novice problem solver and the talented or gifted mathlete. The aha! moment arises when the reader comes up with his or her own elegant solutions other than those provided in this book.

Finally, I would like to thank Heng Ooi Khiang and Lai Chee Chong for correcting some errors and clarifying some fuzzy points, not to say, suggesting alternative solutions to some questions.

Kow-Cheong Yan
kcyan.mathplus@gmail.com

CONTENT

unit
1

What Is a Circle?

People in non-mathematical circles often think that they know what a circle is. Yet, an informal survey among a circle of friends often reveals mixed responses, when asked to define a circle.

Is a circle a completely round shape, like the letter "O," as defined in most dictionaries?

Can you circle the correct answer to the definition of a **mathematical circle**?

        Does a circle consist of

        A.   **only the circumference,**

        B.   **only the area inside the circumference,**

        C.   **both the circumference and the area?**

What about making use of the definition of a *locus* (plural: *loci*), which is the path traced by a set of points that satisfy one or more stated conditions? Or, would drawing a circle 10 centimeters in diameter give away any hint?

If you are still unsure about the correct answer, maybe the following may help:

1. **Cut out a circle. Place one dot anywhere on that circle.**

2. **Draw a circle. Place one dot in the circle, and one dot on it.**

Before you read on, do you see some light regarding a formal definition of a circle?

In the first case, "circle" appears to be the disk of paper that remains after cutting along the circular rim, and "on" seems to suggest *anywhere* on that disk.

In the second case, the word "on" sheds light on any ambiguous meaning of "circle" and confirms that "circle" refers to the line one drew, and not the area contained within it. In terms of the equation of a circle (e.g., $x^2 + y^2 = r^2$), the circle is the ring, and not the inside region (i.e. $x^2 + y^2 < r^2$).

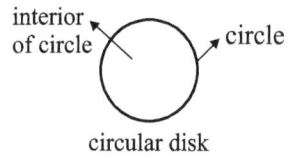

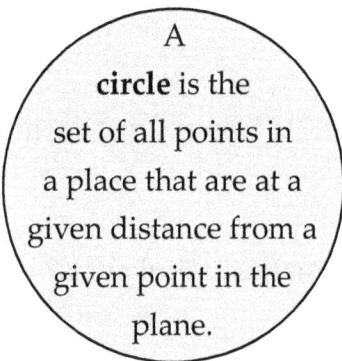

interior
of circle → circle

circular disk

> A **circle** is the set of all points in a place that are at a given distance from a given point in the plane.

So the next time you need to explain a circle to a five-year-old child, you needn't go in circles; simply say that a circle is nothing but an outline defining the shape of a circle. Wait; are you entering some vicious circle?

An Ideal Circle Doesn't Exist

The mathematician's circle has an infinitely thin circumference and a radius that remains constant to infinitely many decimal places, which cannot take any physical form.

If all this circle-talk is making you confused, blame those circles of geometers working in academia. Anyway, after all, no one has seen, nor can anyone draw a perfect circle—I mean an ideal circle, in the Platonic sense—in our imperfect physical world. Of course, this is not mathematics; it is philosophy of mathematics—it is *metamathematics*, if you want a high-tech word, just like *metacognition*, that big-sounding word meaning "thinking about thinking".

Definitions, misconceptions, or confusions about a circle aside, remember that a [perfect] mathematical circle resides in some Platonic realm—a circular hot topic among philosophers—while most mathematicians are content with an imperfect circle, which is good enough for them to make new conjectures (or hypotheses), circular or otherwise.

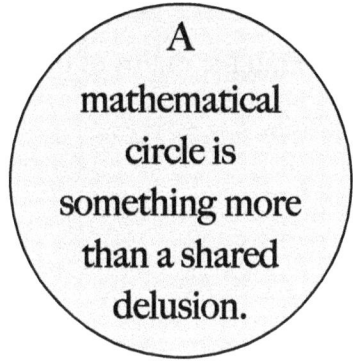

> A mathematical circle is something more than a shared delusion.

In other words, our drawing of a circle on paper is only an approximation of the mathematical, Platonic circle.

> **"The nature of God is a circle of which the centre is everywhere and the circumference is nowhere."**
>
> *Anonymous*

As for the rest of us, we may feel more at home daydreaming about crop circles instead of imagining mathematical circles *à la Plato*, although both are alien to our quasi-circular planet.

Circularly yours

Practice

1. An equilateral triangle is inscribed in a circle and another equilateral triangle is circumscribed about the circle. Find the ratio of the smaller triangle to the bigger triangle.

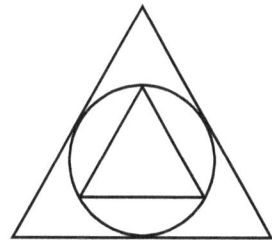

2. A triangle has sides 6, 8, and 10.

    Find the radius of the inscribed circle.

3. Given the unit square, the interior shaded square is formed by joining each vertex of the unit square to the midpoint of its nonadjacent side. What is the shaded area?

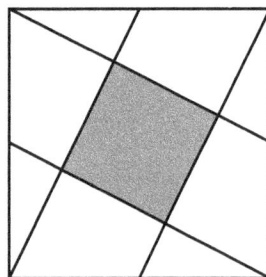

4. Two squares 4 cm by 4 cm overlap, as shown. A vertex of one square lies at the center of the other square. What is the largest possible value of the shaded area?

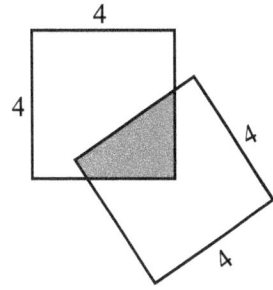

5. A rectangle is divided into four rectangles with areas 15, 45, 108, and x.

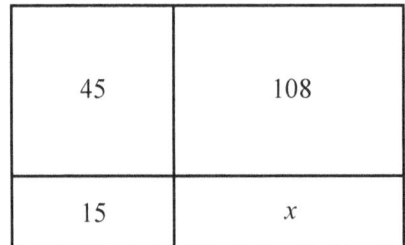

   Find the value of x.

45	108
15	x

6. A rectangle is divided into four
   triangles with areas 16, 24, 25,
   and y.

   Find the value of y.

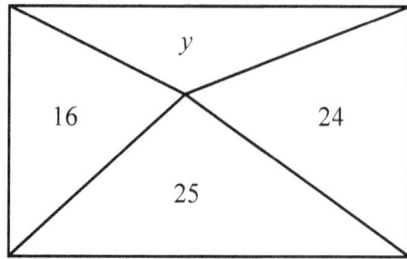

Who first discovered
that the world was
round?

7. A cog-wheel of 12 teeth rotates on its axis round a fixed cog-wheel of 24 teeth.
   How many times does the small cog-wheel make around the big one?

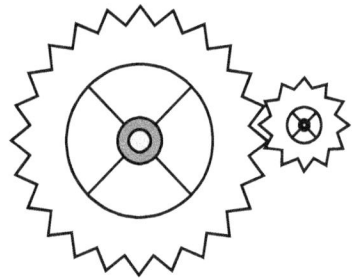

8. Using six slices, Mr. Gan is able to saw a cubical block into 27 smaller cubes, as shown in the figure.

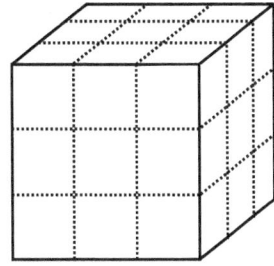

How many slices are needed to saw the block into 64 smaller cubes?

9. The figure shows a cube with corners *A*, *B*, and *C*.

Calculate the measure of angle $\angle ABC$.

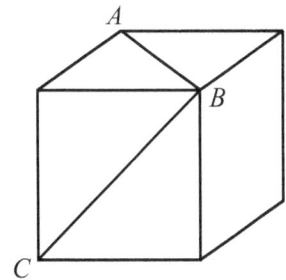

10. Find the area of the trapezoid, as shown in the figure.

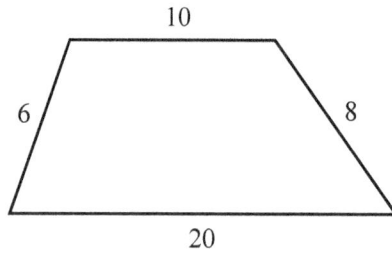

U.S.	Singapore
trapezium	***trapezoid***
trapezoid	**trapezium**

A trapezium in Singapore is known as a trapezoid in the United States.

Further reading

Bryant, J. & Sangwin, C. (2008). *How round is your circle?* Princeton: Princeton University Press.

unit 2

Three Famous (or *Notorious*) Geometric Problems

Three notorious geometry problems that had bedeviled ancient Greek mathematicians for centuries were: "trisecting an angle," "duplicating a cube," and "squaring the circle."

Square the Circle

The antique geometry problem of squaring the circle goes like this: *Given a circle in the plane, is it possible to construct a square of the same area, using only an unmarked ruler and a compass?*

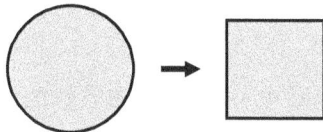

Attempts to solve this problem as far back as 460 B.C. had defeated generations of geometers. The proof that it is impossible to "square the circle" came only some 2000 years later.

Duplicate a Cube

Legend tells that there was a plague in Athens, and the inhabitants asked the oracle of the god Apollo for help. The oracle replied that the plague would cease if the altar to Apollo were exactly doubled in size. In other words, the Athenians would need to construct an altar whose volume is twice the original one—they need to duplicate a cube.

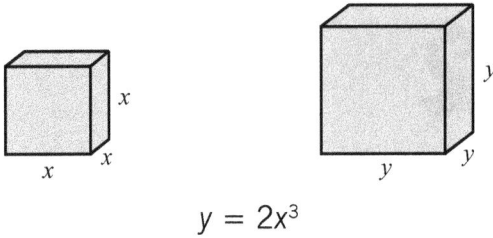

$$y = 2x^3$$

So, if the altar was a geometric cube of edge length one cubit, Apollo would only be pleased by a cube whose edge length was $\sqrt[3]{2}$ cubits.

> **Trisect an angle**: divide an angle into three, with compasses and an unmarked ruler (or straightedge)
>
> **Duplicate a cube**: form a cube whose volume is double another cube
>
> **Square the circle**: form a square with the same area of a circle, with compasses and an unmarked ruler

Trisect an Angle

The other geometric problem that had humbled the Greeks was the trisection of the angle.

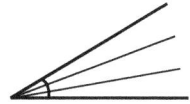

The Greek geometers knew how to bisect a line and also how to trisect it, or divide it into any number of equal parts. They also knew how to bisect an angle, and by repeating this they could divide it into four parts, or further divide it into $2 \times 2 \times 2 \times \cdots \times 2$ (or 2^n) equal parts, but they could not find a ruler and compare construction for trisecting an arbitrary angle.

The solutions of these geometric impossibilities were resolved not by geometric, but by algebraic means. For example, the proof that π is a non-algebraic (or transcendental) number (i.e., pi cannot be the root or solution of a polynomial) leads to the impossibility of constructing a square with the same area of a circle.

Every now and then, circle squarers, cube duplicators, or angle trisectors do show up, denying that their proofs are flawed. These "mathematical cranks" refuse to back down, often denouncing the mathematical brethren of conspiracy in admitting their ignorance.

A mathematical crank has zero humility and infinite pride.

Practice

1. The larger circle circumscribes a square, which circumscribes a smaller circle. Find the ratio of the area of the smaller circle to the area of the larger circle.

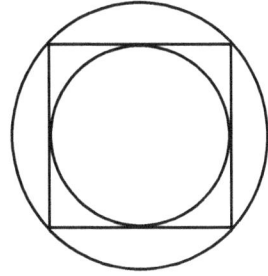

2. The perimeter of the smaller square is 68 cm.

   If the area surrounding it has area 111 cm^2, find the length of the larger square.

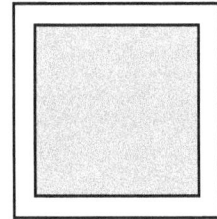

3. All cross-sectional areas of the solid are congruent equilateral triangles. If the volume of the solid is 75 cm³, find the perimeter of the cross-section.

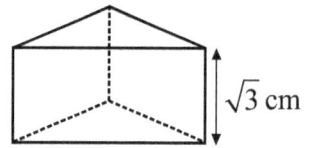

$\sqrt{3}$ cm

4. Given that the two triangles are similar, what is the ratio of the area of the large triangle to the area of the small triangle?

$\sqrt{3}$ $\sqrt{51}$

5. Given that $XY = 14$ cm, find the area of the shaded region, expressing your answer in terms of π.

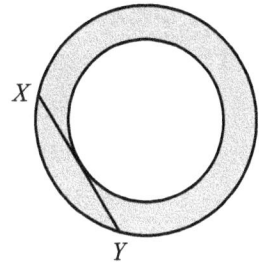

6. P is a square. How would you double its area, and yet have the same square shape?

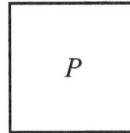

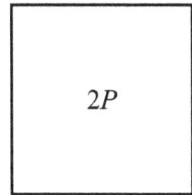

7. A circle of radius 10 cm intersects another circle of radius 15 cm at right angles. What is the difference of the areas of the unshaded regions, leaving your answer in terms of π?

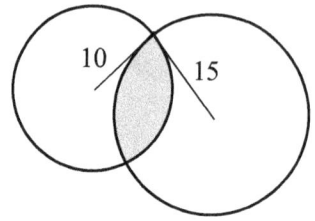

8. How many triangles are there in this figure?

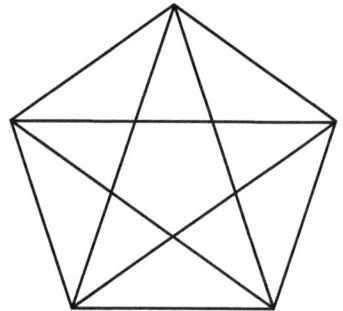

9. A square contains two quarter circles, as drawn in the figure.

Find the difference between the two shaded areas.

(Take π to be $\dfrac{22}{7}$)

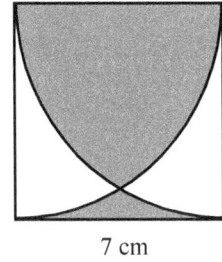

7 cm

10. What fraction of the figure is shaded?

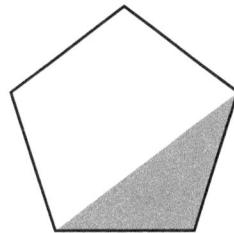

Given a long, rectangular strip of paper, how would you make a *regular pentagon* (a five-sided figure with all sides the same length and all angles the same size)?

unit
3

◇◇

Non-Euclidean Geometry
for *Goondus**

Explaining *Non-Euclidean Geometry* to a layperson is like sharing God's Love to an atheist or agnostic, simply because the results of non-Euclidean geometry are repugnant to common sense—they are counterintuitive to most Euclidean geometry (named after the Greek geometer, Euclid) taught in elementary school.

Traditionally, "Non-Euclidean Geometry" is taught at the university level, but these days it is not uncommon for it to be introduced as an enrichment topic among middle school mathletes, or to be given as a topic for a math project. Even among undergraduates, being exposed to non-Euclidean geometry does not generally arouse much excitement or interest before, during, or after the few lectures covering its counter-intuitive results. This is simply because the results are often parroted in the most boring, sterile manner, devoid of the controversies arising from developing such geometries.

Teaching the gist of non-Euclidean geometry may be likened to "explaining the unexplainable" to innocent minds. So let me try to do an impossible task, by attempting an inaccurate *goondu*-style explanation of what non-Euclidean geometry may mean to the lay public (or for the rest of us)—who may have, at best, some half-baked idea what Euclidean geometry is all about. Unknown to most educated mortals, non-Euclidean thinking (if we know its meaning and practice it) is a prerequisite to creativity and innovation in any discipline or field.

*Goondus are the Asian equivalent of idiots, dummies, or morons.

A [Very Short] History of Non-Euclidean Geometry

Once upon a time, there was a Greek mathematician (or plagiarizer, as some of his critics want us to believe) named Euclid (*c.* 300 BC) who compiled some 400 odd theorems, and some axioms and postulates—those undefined terms (such as *points*, *lines*, and *planes*) to be taken on faith—on plane and solid geometry, and on number theory ("higher arithmetic") into 13 books. Euclid's *Elements* was known as a mathematicians's bible for ages, until some French mavericks conspired to throw Euclid out of the mathematics classroom.

Prof. Euclid's postulates such as "Two points define a line" and "Two right angles are equal" did not raise any suspicion, except for the "parallel postulate" which says that *given a line and a point not on it, there exists only one line that passes through the point and never intersects the given line*. In street jargon, *two parallel lines never meet*.

Euclid
(c. 325 BC–265 BC)
Greek mathematician

> **There is no royal road to geometry.**
>
> — Euclid

Since mathematicians (in this case, geometricians, as they were then known) refuse to take anything for granted, it is no surprise that doubters of Euclid's axioms were uneasy about his self-evident parallel postulate, and thus toyed with the idea whether it could be derived from the other basic axioms (self-evident assumptions).

The Troublesome Axiom

Attempts to prove or disprove Euclid's troublesome axiom had bewitched generations of mathematicians; in fact, today, we know that no such proof exists. For example, Gerolamo Saccheri, a Jesuit priest and professor, tried an indirect proof, by assuming that the parallel axiom was false; and derived a large body of results. He believed, erroneously, that he had *proved* the parallel axiom, but failed to realize that his results were theorems in a new *non-Euclidean geometry*.

Then, in the first half of the 19th century, the Russian Nicolai Lobachevsky and the Hungarian Janos Bolyai independently used strategies similar to Saccheri, but with a twist. They replaced the parallel axiom with the axiom that given a line and point not on the line, there exist, not just one, but many lines containing the given point and not meeting the given line.

Although many non-Euclidean geometries had been formulated to challenge Eucliean geometry, two species of non-Euclidean geometry prevail today:

Hyperbolic geometry (infinite number of ultra-parallels): Given a line *l* and point *P* not on *l*, every line, coplanar with the line *l* and containing the point *P*, meets *l* in exactly one point.

Elliptic geometry (no parallels): Given a line *l* and point *P* not on *l*, there exists more than one line, coplanar with line *l*, containing the point *P*, that does not meet *l*.

In layman terms, it means that two parallel lines *can* meet; or that when they do not, there are many lines passing through a given point that are parallel to a given line. Do not worry too much if those two results sound nonsensical!

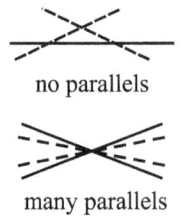

no parallels

many parallels

Aftermath of Non-Euclidean Geometry

What is amazing is that non-Euclidean geometry is more than the mere denial of Euclidean geometry; it cannot contain contradictions unless ordinary Euclidean geometry also has contradictions. For instance, in Euclidean geometry, the sum of the interior angles in a triangle is 180 degrees, but in a non-Euclidean plane, the sum may be less or more than 180 degrees. Imagine drawing a triangle on the surface of a sphere, and convince yourself that the sum exceeds two right angles.

The sum of a triangle varies according to the type of geometry:

= 180° — Euclidean geometry

< 180° — Lobacheviskian geometry

> 180° — Riemannian geometry

Bluntly speaking, non-Euclidean geometries point to the fact that there is nothing "real" about the mathematical universe. Besides, these new non-Euclidean geometries were as "good" as Euclidean geometry, in which all of the axioms of Euclid, except the parallel postulate, were true.

The ratio of the circumference, C, of a circle to its diameter, d, depends on the type of geometry.

In Euclidean geometry, $\dfrac{C}{d} = \pi$

In Lobachevskian geometry, $\dfrac{C}{d} > \pi$

In Riemannian geometry, $\dfrac{C}{d} < \pi$

Interestingly, many natural phenomena favor a non-Euclidean explanation for their occurrences and successes. For instance, without Riemannian non-Euclidean geometry, Einstein would not be able to formulate his theory of general relativity. Indeed, non-Euclidean geometries dealt a blow to the edifice of Euclid, just like the discovery of irrational numbers (like the square roots of two or the number pi) some two centuries earlier (*c*. 500 BC) shattered Pythagoras' dream that the workings of the universe could be explained in rational terms—in whole numbers and their ratios.

Non-Euclidean Dialect, Anyone?

Today, the term "non-Euclidean" is employed in non-mathematical fields as a label for unconventional, non-traditional, or radical thinking. For example, an organization without a non-Euclidean outlook in embracing chaos and unorthodox views from its staff is unlikely to be creative and innovative. Unquestioned (or blind) Euclidean obedience or submission to the management would only lead to frustration and eventually to non-Euclidean action.

The discoveries of non-Euclidean geometry were like the biblical king Saul who was looking for some donkeys and found a kingdom. Or, like Columbus who was looking for some new land but found a continent.

Non-Euclidean Thinking for the 21st Century

It is a pity that tens of millions of adults continue to leave school today, without the slightest hint what Euclidean and non-Euclidean geometries are all about, and how non-Euclidean thinking helped revolutionize the world of art and of science in early centuries: cubism, communism, relativity, quantum mechanics, or Darwinism.

And in our time, would iPhone, wireless e-mail, Google Earth, and the like qualify as non-Euclidean products of technology? Most importantly, are you ready to think and speak non-Euclideanly?

Practice

1. A sphere of radius 1 meter is rolled into the corner of a room, touching a smaller sphere. Find the radius of the smaller sphere that is tangent to the same two corners of the walls and touches the larger sphere.

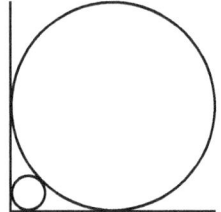

2. The diagram shows a right-angled hollow pipe of diameter 20 cm. Given that its ends are open, find the external surface area of the pipe. (Take $\pi = 3.14$)

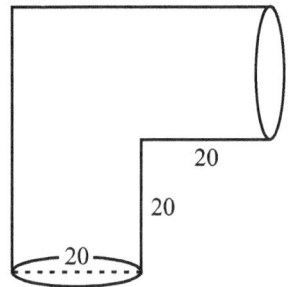

3. The radii of the faces of a frustum of a cone are 3 cm and 2 cm and the thickness of the frustum is 3.6 cm.

Determine the volume of the frustum.

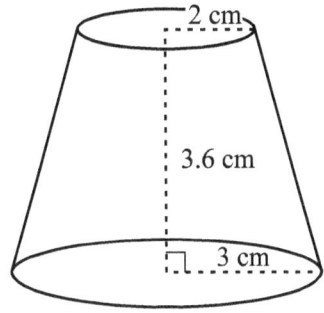

4. Without using trigonometry, show that angle C in the figure, made up of three squares, equals the sum of angles A and B.

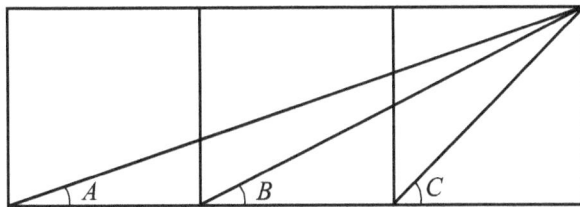

5. What fraction of the following figures are shaded?

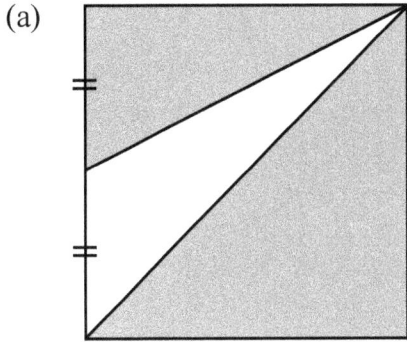

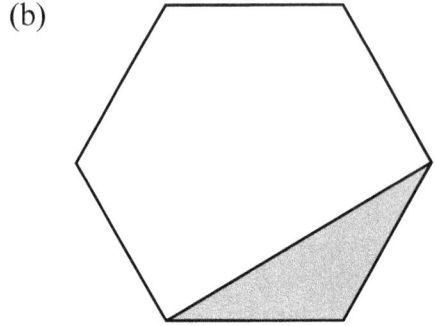

(a)

(b)

It's a regular hexagon.

6. How long (in feet) is the spiral on the cylindrical column?

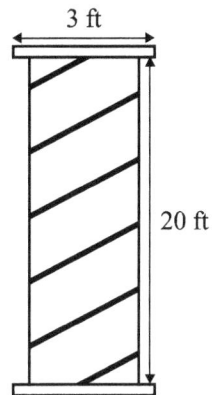

3 ft

20 ft

7. Mr. Yan intends to have a circular pool in his triangular plot of land, as shown. Find the radius of the largest pool that could fit.

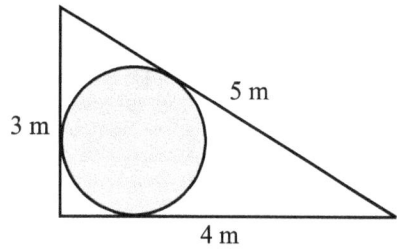

3 m

5 m

4 m

A circle is a figure with 0 corners and only one side.

8. A regular hexagon (6-sided figure) and an equilateral triangle have the same perimeter. Find the ratio of their areas.

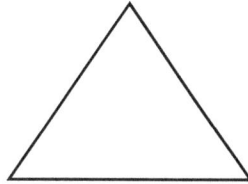

9. The area of a rectangle of length and width, expressed in whole numbers, is 96 cm^2, and the perimeter is 40 cm. What is the length of the rectangle?

The earth makes a resolution every 24 hours.

10. The figure given represents a cuboid. The areas of three touching faces are 12 cm^2, 15 cm^2, and 20 cm^2. Determine the volume of the cuboid.

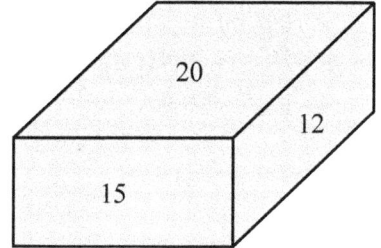

20

12

15

unit

4

Regions of a Circle

Study the sequence of circles.

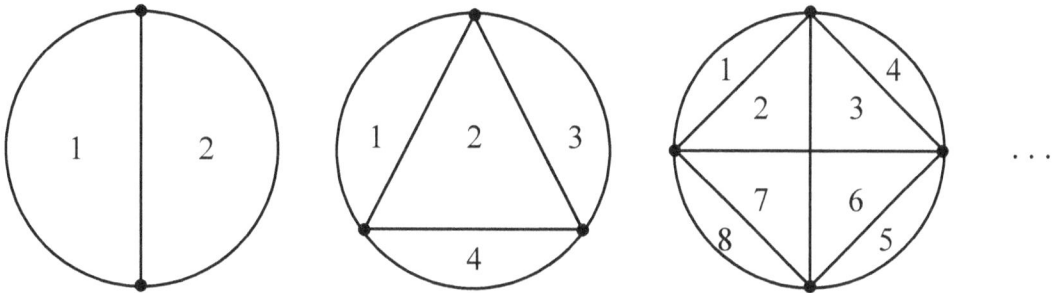

In the first circle, two points joined by a straight line segment yields two regions. In the second circle, three points joined by three line segments produces four regions, and in the third circle, four points by four line segments yields 8 regions.

Continuing in this manner, what will be the greatest number of regions in the fourth and fifth circles? Construct the circles and convince yourself of the results.

Number of points on the circle	Number of regions into which the circle is divided
1	1
2	2
3	4
4	8
5	?

A Surprising Result

The 4th circle can be divided into 16 regions, but the 5th circle does not divide into 32 regions. In fact, it has been proved that the 5th circle can be divided into 31 regions. Check it out!

The general formula for n points on the circle can be divided into

$$\frac{n^4 - 6n^3 + 23n^2 - 18n + 24}{24} \text{ regions,} *$$

and not 2^{n-1} regions, as is the case for $n = 1, 2, 3, 4$.

So, beware of formulating any result only from the first few terms. Do not trust your intuition!

* In combinatorial terms, the number of regions is given by

Number of region $= 1 +$ number of lines $+$ number of intersections

$$= 1 + \binom{n}{2} + \binom{n}{4} \text{ for } n \geq 4$$

$$= 1 + \frac{n(n-1)}{1 \times 2} + \frac{n(n-1)(n-2)(n-3)}{1 \times 2 \times 3 \times 4}$$

Practice

1.   A cube is made up of 6 sides. How many cubes will 138 sides make up?

To cube something is to raise it to the power of three.

The result of cubing is a *cube number*:

$1^3 = 1, 2^3 = 8, 3^3 = 27, ...$

2. The figure shows the vertices of a triangle with sides 6 cm, 8 cm, and 10 cm coinciding with the centers of three mutually tangent circles.
Calculate the radius of the smallest circle.

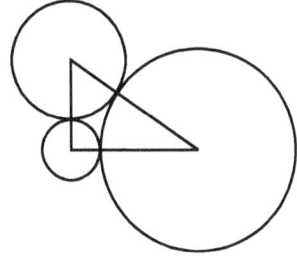

3. A small circle is inscribed inside a semicircle.

Find the ratio of the area of the circle to the area of the shaded region.

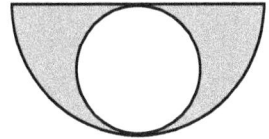

4. The figure shows semicircles drawn on each side of a right triangle with areas 9π, 16π, and 25π. Find the area of the triangle.

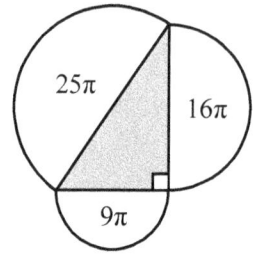

25π

16π

9π

5. The area of the circle is 20π cm^2. Find the area of the square.

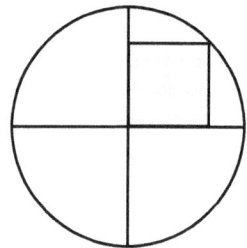

There are no missing data!

6. The figure shows a succession of squares within circles, each touching the other. If the diameter of the outside circle is 10 cm, what is the diameter of the inside circle?

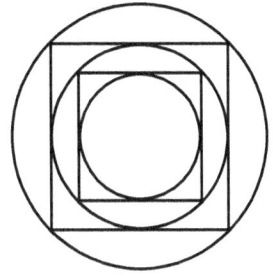

7. In the diagram, the diagonals of the trapezoid divide the trapezoid into four triangles. Given that the areas of the two shaded triangles are 18 cm² and 32 cm², find the area of the trapezoid.

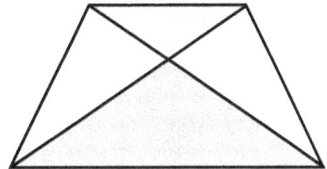

8. In the figure, D is the midpoint of AC.

    Show that

    (a) area $\triangle AED$ = area $\triangle CED$,

    (b) area $\triangle BEA$ = area $\triangle BEC$.

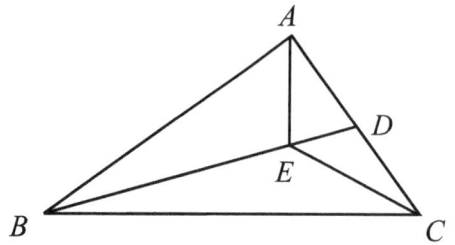

9. By revolving a 3-4-5 right triangle about one of its sides, a solid may be generated. Each side produces a different solid. By considering the two solids about the two shorter lengths, find the ratio of the smaller volume to the bigger volume.

10. A plot of land of area 25 km² is represented on the map by 4 cm². Find the scale of the map.

Have you heard of the four-color map problem?

Yes, on a map, only four colors are enough to ensure that no neighbouring countries are colored the same. But so far only a computer proof exists! Why would God allow such a beautiful theorem to have such an ugly proof?

unit 5

<hr />

THAT *HOLY* LITTLE GEOMETRY BOOK

Biographical sketches of great mathematicians suggest they were hooked to mathematics because of the enchanting material of Euclidean Geometry—named after the Greek mathematician Euclid (c. 300 BC), who compiled 13 books of geometry entitled *The Elements*. In fact, many great minds in mathematics professed that they religiously and painstakingly went through it, proposition by proposition.

And we are also told that Euclid's *Elements* was the bestseller of all times, except for the Holy Bible, going by numerous revisions before Euclidean geometry was no longer the staple of high school mathematics. So, in many ways, Euclid's *Elements* was like a mathematician's bible.

> "If the triangles were to make a God they would give him three sides."
>
> Montesquieu (1689-1755)
> French political philosopher

Intimacy with Euclidean Geometry

Even young Albert Einstein was aroused by a small textbook on Euclidean geometry, given by an uncle, when he was in his twelfth year. In his *Autobiographical Notes*, the physicist wrote of "the holy little geometry book." He was enticed by assertions such as the three altitudes (or perpendicular heights) of a triangle intersect in one point, although not evident, could be proved without any shred of doubt.

British philosopher and mathematician Bertrand Russell's intimacy with Euclid's *Elements* happened when he was eleven, with his 18-year-old brother tutoring him. In his *Autobiography*, Russel says: "This was one of the great events of my life, as dazzling as first love. I had not imagined there was anything so delicious in the world. From that moment until I was thirty-eight, mathematics was my chief interest and my chief source of happiness."

The historian of mathematics, Howard Eves, too, was lured by Euclid's *Elements*, which contained a large number of geometric facts all presumably deduced from a small handful of very readily accepted initial assumptions (known as *axioms*).

Dr. Eves remarked that unlike high school algebra which is largely a collection of procedures, Euclidean geometry contains the essential mathematical ingredient of deduction—the process of starting with the concrete and proving the general. It is not until one encounters abstract algebra in university that algebra becomes genuine mathematics, because up to high school, algebra primarily involves the manipulation of symbols.

Euclid's *Elements* at a glance

Beginning with a list of 23 definitions, 5 postulates, and 5 common notions, Euclid's *Elements* is a compilation of 13 books consisting of 465 theorems on plane and solid geometry, on elementary number theory, on geometric algebra, and on the theory of proportion and incommensurable line segments.

Over a thousand editions of Euclid's *Elements* have appeared since the first one printed in 1482, and for more than two millennia this work has dominated all teaching of geometry.

How to Spot a Potential Math Student

According to Dr. Eves, whether a college student will do well in mathematics is to ask the student how he or she did in high school geometry and algebra. If the reply is: "I did quite well in algebra, but poorly in geometry," then that student probably *is not* a potential mathematics student. On the other hand, if the reply is: "I did really well in geometry, but algebra held much less interest for me, then that student probably is a potential mathematics student."

What Eves is saying is that exposure to the rigors of Euclidean geometry proves a better indicator of one's mathematical potential to read advanced mathematics in university.

To the average person, those old books on advanced Euclidean geometry, lying in dusty storerooms, would not arouse much intellectual excitement. Like fine wine, geometry *à la Euclid* is an acquired taste. There are fine French wines, fine Italian wines, fine German wines, and so forth. Poncelet, Gergonne, Brianchon, Beltrami, Klein, and company—such old-time geometers.

The goddess Mathesis

How ever young and beautiful you are

I desire to share your beauty and utility

To extol your attractions and paradoxes

So that others too may be won over

Practice

1.   What fraction represents the shaded part of the regular hexagon?

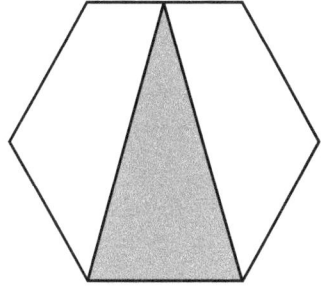

2.   The diameter of the inner circle is 4 cm and of the outer circle is 6 cm.

Which is smaller: the area of the inside circle, or the area between them?

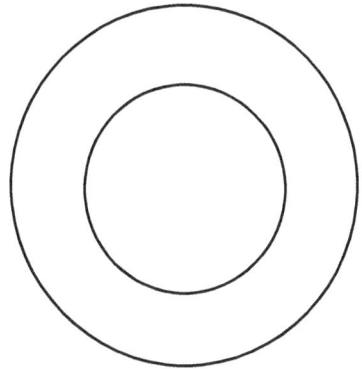

3. Four quarters of a circle of radius 2 cm are arranged to make the shape below. What is its area?

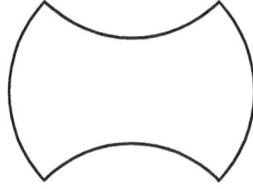

4. In the parallelogram, the angles at A and B are trisected.

What is the measure of angle MCN?

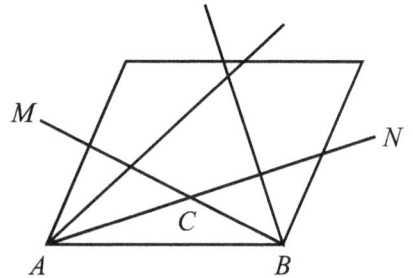

5. Calculate the shaded area in the figure on the right.

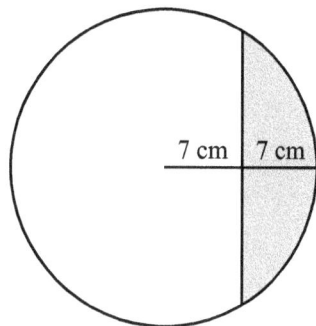

---

6. What fraction represents the shaded part of the regular octagon?

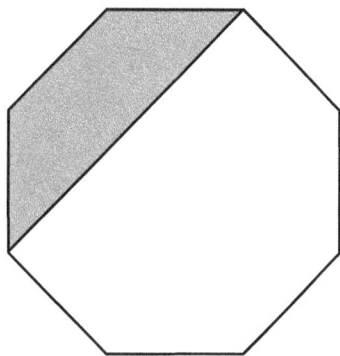

7. Find the radius of the circle in the figure.

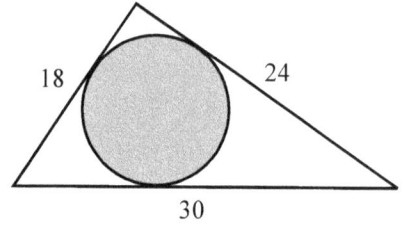

8. How many triangles are there in the figure?

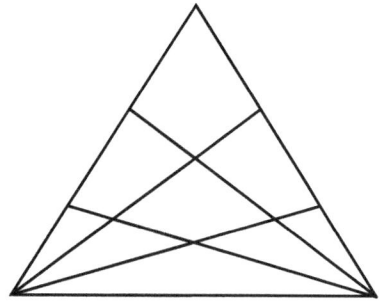

9. The figure depicts an inverted cone of height H and radius r.

The cone contains water to a depth of $\frac{1}{2}H$.

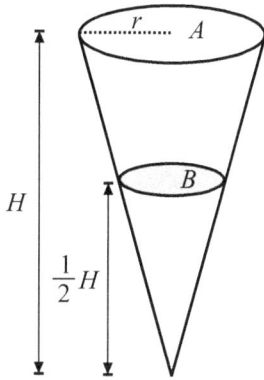

Figure 1

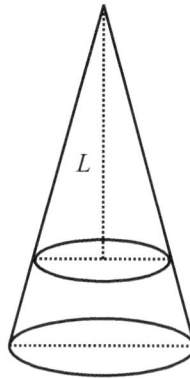

Figure 2

(a) Calculate the value of $\dfrac{\text{Surface area } A}{\text{Surface area } B}$.

(b) Find the volume of water in the cone if it can hold 480 cm³ of water when full.

(c)   The cone is inverted with the water resting on the circular base of the cone, as shown in Figure 2. Express L, the distance from the tip of the cone to the water surface, in terms of H.

10.   The diameters in the diagram are 6 cm, 4 cm, 4 cm, and 2 cm.

Which is larger: the lighter shaded area or the darker shaded area?

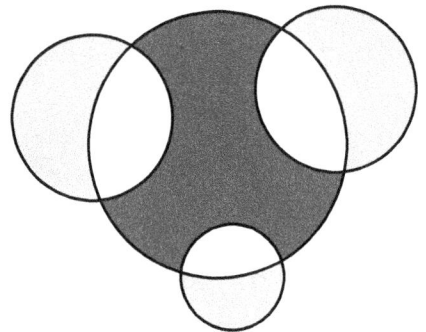

unit 6

Fun with Areas and Perimeters

A given perimeter can have many different areas, and a given area can be enclosed by many different perimeters.

For example, rectangles of perimeter 16 cm may have different areas.

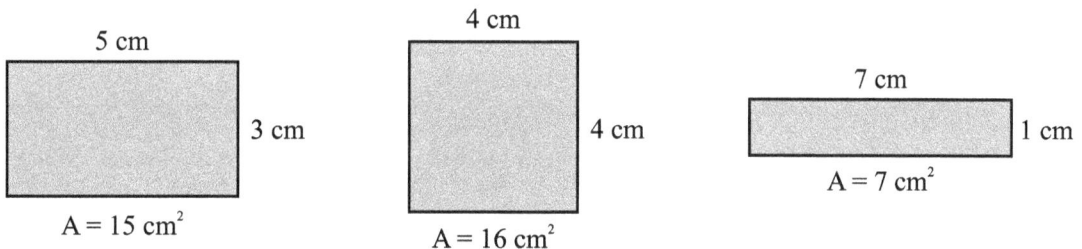

5 cm
3 cm
$A = 15$ cm^2

4 cm
4 cm
$A = 16$ cm^2

7 cm
1 cm
$A = 7$ cm^2

Each rectangle yields different areas, although the perimeter in each case is the same.

Similarly, rectangles of area 12 cm^2 may have different perimeters.

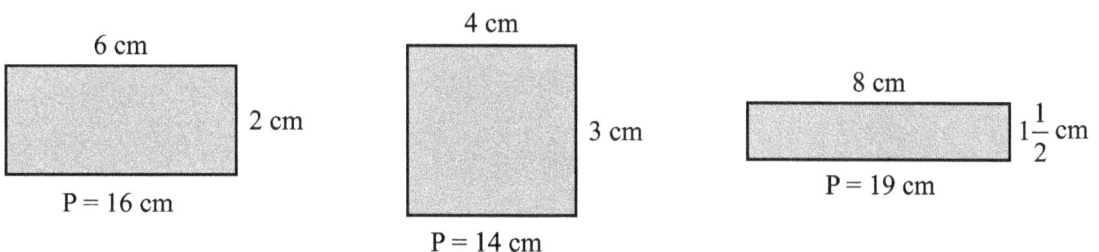

6 cm
2 cm
$P = 16$ cm

4 cm
3 cm
$P = 14$ cm

8 cm
$1\frac{1}{2}$ cm
$P = 19$ cm

Each rectangle yields different perimeters, although the area remains unchanged.

1. Given a fixed perimeter, what kind of rectangle will yield the greatest area?

What if the shape of the figure needn't be quadrilateral?

2. Given a fixed area, what kind of rectangle will yield the least perimeter?

What if the figure isn't restricted to four sides?

Worked Example 1

(a)   Given four strings of equal length.

   (i)   Form a circle from the first piece of string.

   (ii) Cut the second piece of string into two equal parts and form two equal circles.

   (iii) Cut the third piece of string into three equal pieces and form three equal circles.

   (iv) Cut the fourth piece of string into four equal pieces and form four equal circles.

(b)   Show that the sum of the circumferences of each group of similar circles is the same.

(c)   Compare the sum of the areas in each group. What can you conclude?

Solution

The more circles formed with the same total length of string, the smaller the total area of the circles.

Circle	Diameter	Circumference of each circle	Sum of circles' circumferences	Area of each circle	Sum of circles' areas
P	12	12π	12π	36π	36π
Q	6	6π	12π	9π	18π
R	4	4π	12π	4π	12π
S	3	3π	12π	2.25π	9π

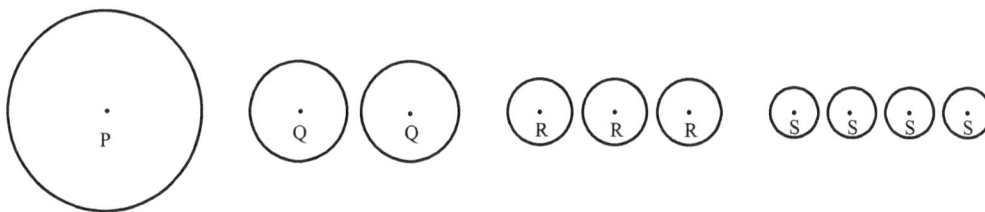

Explanation

When 2 equal circles were formed, the total area of the two circles was one-half that of the large circle.

Similarly, when four equal circles were formed, the total area of the four circles was one-fourth of the area of the large circle.

This seems to go against one's intuition. Yet if we consider a more extreme case, with 100 smaller equal circles, we would see that the area of each circle becomes extremely small and the sum of the areas of these 100 circles is one-hundredth of the area of the larger circle.

Can you explain this rather disconcerting concept?

Practice

1. What is the maximum number of rectangular blocks, each measuring 3 cm by 5 cm by 7 cm, that can be placed inside a rectangular box measuring 14 cm by 15 cm by 16 cm?

2. If three sides of a triangle are all whole numbers, and exactly two of the sides are equal, what is the least possible perimeter of the triangle?

3. Look at the two triangles below. Which has the larger area?

Hint: Try cutting one triangle up to make the other triangle.
    There are of course other methods.

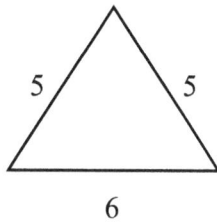

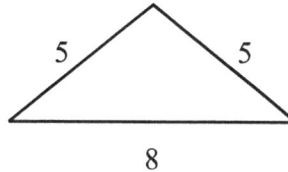

4. Find the area of a regular 12-sided polygon which is inscribed in a circle of radius r units.

PC Geometry

Vertically challenged (short)

Horizontally challenged (fat)

5. If the length of a diagonal of a square is $x^2 + y^2$, find the area of the square.

---

6. What is the smallest rectangular area of wrapping paper needed to wrap the solid in the diagram?

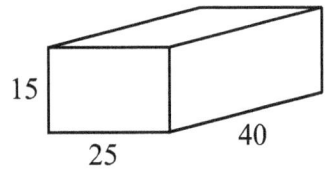

15

25

40

7. How many circular pipes with an inner diameter of 1 cm are needed to carry the same amount of water as a pipe with an inner diameter of 6 cm?

8. In the figure, the radius of the circle is 1 cm and the width of the rectangle is 1 cm. Find the shaded area.

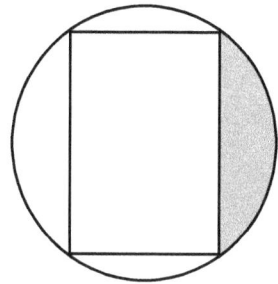

9. Two beams are supported vertically between two vertically walls.
    The ends of the beams are 6 m and 4 m above the ground level.
    Find the height of the point where they cross, X, above the ground level.

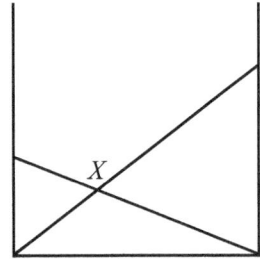

---

10. How many diagonals can be drawn in a 100-sided polygon?

• Solve a simpler problem.
• Look for a pattern.

unit
7

Always a
Parallelogram!

Draw a few quadrilaterals of any shapes. Then locate the midpoints of the four sides of each quadrilateral. Join these points consecutively. Do you get a parallelogram inside each quadrilateral each time? How and why did this happen?

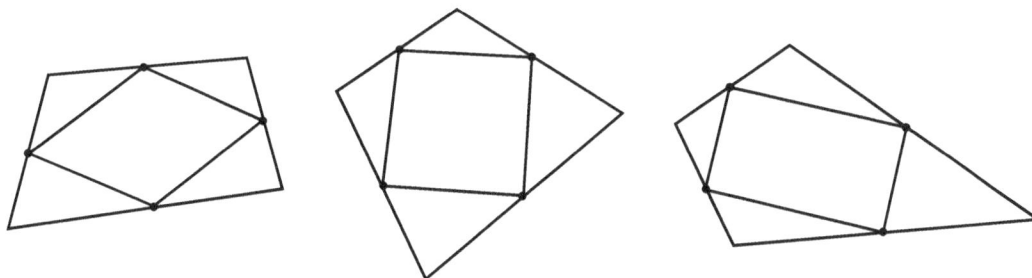

Worked Example 1

Investigate what shape the original quadrilateral should be so that the inside parallelogram is

(a) a rectangle,

(b) a square,

(c) a rhombus.

Solution

(a) When the diagonals of the original quadrilateral are perpendicular, the parallelogram is a rectangle.

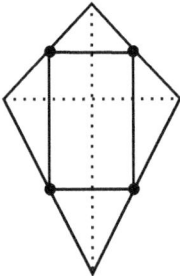

(b) When the diagonals of the original quadrilateral are congruent (or are equal in length) and perpendicular, then the parallelogram is a square.

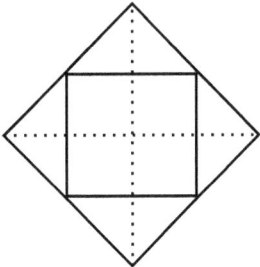

(c) When the diagonals of the original quadrilateral are congruent, then the parallelogram is a rhombus.

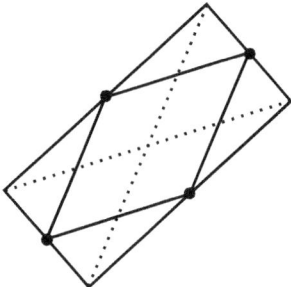

The proof rests on a simple theorem that states that a line segment joining the midpoints of two sides of a triangle is parallel to and half the length of the third side of the triangle.

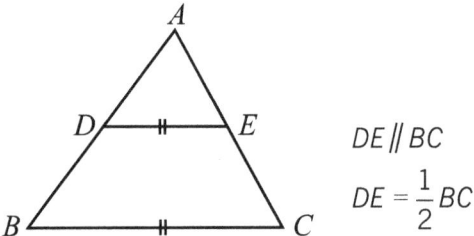

$DE \parallel BC$

$DE = \dfrac{1}{2} BC$

In $\triangle ADB$, the midpoints of sides AD and AB are F and G respectively.

$$FG \parallel DB \text{ and } FG = \frac{1}{2}BD$$

and

$$EH \parallel DB \text{ and } EH = \frac{1}{2}BD$$

$\therefore \quad FG \parallel EH$ and $FG = EH$.

This establishes $EFGH$ as a parallelogram.

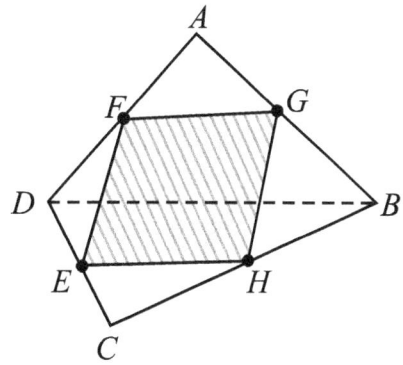

Furthermore, if the diagonals DB and AC are congruent, then the sides of the parallelogram must also be congruent, since they are each one-half the length of the diagonals of the original quadrilateral. This results in a rhombus.

Similarly, if the diagonals of the original quadrilateral are perpendicular and congruent, then since the sides of the parallelogram are in pairs, parallel to the diagonals and half their length, the adjacent sides of the parallelogram must be perpendicular and congruent to each other, making it a square.

Practice

1. The figure shows a square defined by four dots.

    Move two dots to create a square twice as big as the one defined by the dots as they are presently arranged.

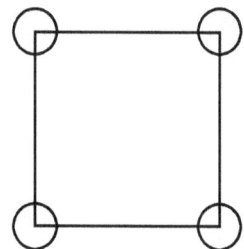

2. Without using Heron's formula, find the area of the triangle.

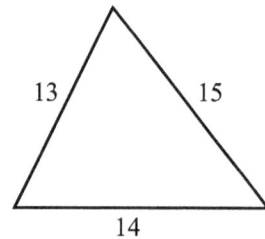

3. The interior angle of a regular polygon is 140°. How many sides does the polygon have?

4. How many rectangles are there in the figure?

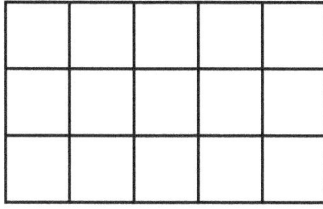

5. A cube has its total surface area numerically equal to its volume.

Find the volume of the cube.

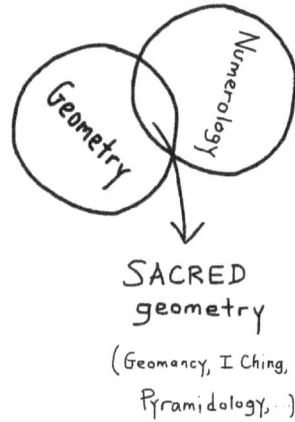

Thou shalt not dabble in pseudoscience.

Geometry

Numerology

SACRED geometry

(Geomancy, I Ching, Pyramidology, ...)

6.   The total surface area of the solid is 192 cm². Find its volume.

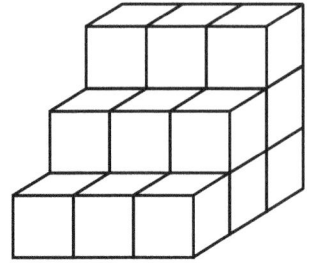

7.   Points *A*, *B*, *C*, *D*, and *E* lie on the circle.

   Find the sum of the interior angles *A*, *B*, *C*, *D*, and *E*.

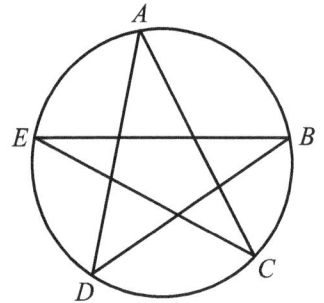

8. *ABCD* is a unit square.

    Find the area of region *ADEF*.

    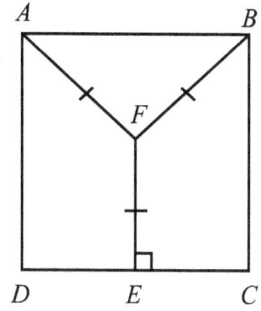

9. A circular track is formed by circles *A* and *B*.

    The diameter of circle *C* is a chord of circle *A* and is tangent to circle *B*. What is the relationship between the circular track and circle *C*?

    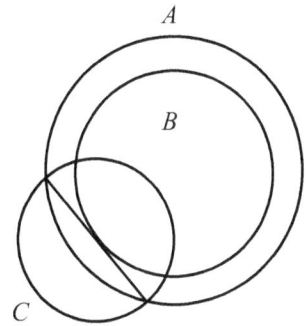

10. A figure is created by beginning with a 2 × 2 square and continually adding to the figure as shown below.

What is the area of the figure?

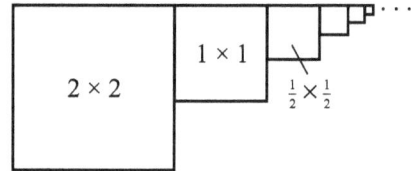

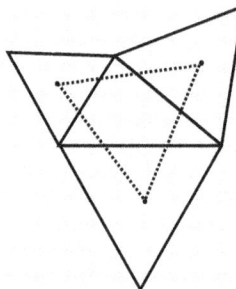

unit 8

The Malfatti's Problem

Given a triangle, how do you construct three non-overlapping circles inside it so that their total area is as large as possible?

The "packing" problem was first posed in 1803 by the Italian mathematician Gianfrancesco Malfatti (1731–1807) who thought he knew the answer: Choose the circles in such as way that each one touches two sides of the triangle and both the other circles.

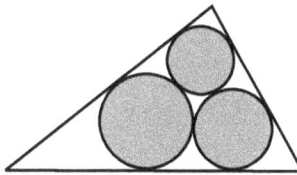

Over a century mathematicians thought the problem had been solved. However, in 1930, somebody noticed something very strange: in the particular case of an equilateral triangle, Malfatti's "solution" is not correct. In his configuration:

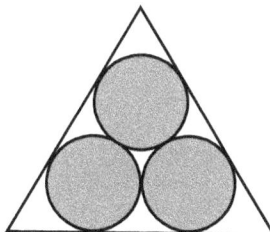

The circles occupy a fraction

$$\frac{\pi\sqrt{3}}{(1+\sqrt{3})^2} \approx 0.729$$

of the triangle's area, but you can do slightly better by using the biggest possible circle and two smaller ones:

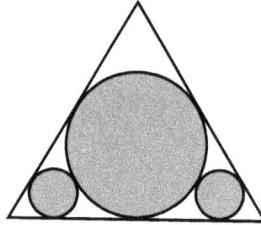

because the fraction then turns out to be

$$\frac{11\pi}{27\sqrt{3}} \approx 0.739$$

And, in 1965, the historian-mathematician, Howard Eves, noticed something strange still—if the triangle in question is long and thin, Malfatti's solution

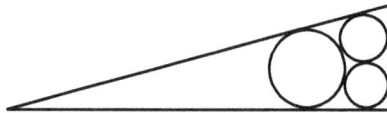

is rather obviously not correct.

A better result if we choose the circles as follows.

Finally, in 1967, Michael Goldbery demonstrated that Malfatti's "solution" is never correct, whatever the shape of the triangle.

The correct solution always has one of the forms below, with one of the circles touching all three sides.

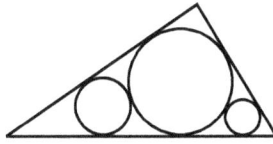

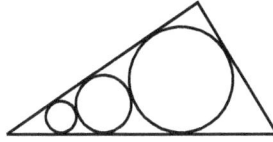

And, in 1992, a complete solution to the Malfatti's problem of arranging three non-overlapping circles of greatest total area in a triangle was given by V. A. Zalgaller and G. A. Los.

Practice

1. One side of a triangle is 30 cm long and another is 40 cm long. How long must the third side be so that the area of the triangle is maximum?

2. A bar of chocolate with 18 "chunks" is to be divided among 18 children. What is the minimum number of "snaps" that you have to make to have 18 individual chunks?

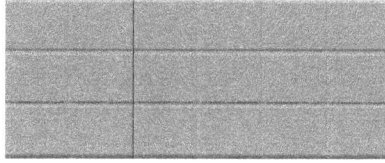

3. The diagram shows two gears. Gear A has 32 teeth and gear B has 16 teeth.

    Taking negative numbers to denote clockwise turns and positive numbers for counterclockwise turns, find

    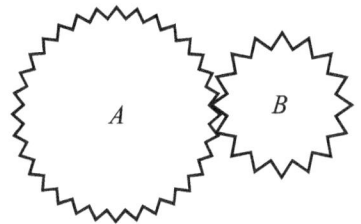

    (a) the number of revolutions for B if A is turned +8 revolutions,

    (b) the number of revolutions for A if B is turned −12 revolutions.

4. A one-meter cubic box is placed flat against a wall. A ladder $\sqrt{15}$ m is positioned in such a way that it touches the wall as well as the free horizontal edge of the box. At what height does the ladder touch the wall?

5. Find the area of an isosceles right triangle whose perimeter is $2a$.

A devil's definition

Geometry: The worst man-made subject ever that torments high school students, in particular. Like algebra word problems, geometric proofs are murderous.

Geometry was the past time of ancient Greek philosophers who relied on slave labor to give them leisure time to construct their proofs.

6. The area of a circle inscribed in an equilateral triangle is 48π. Find the perimeter of the triangle.

7. Find the ratio of the diagonal of a regular pentagon to its side.

8. A cylindrical hole 6 cm long is drilled through the center of a solid sphere. What is the volume that remains?

---

9. In the figure, $AB = AC = BD$. Express q in terms of p.

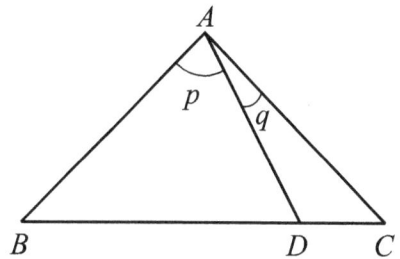

10. In the figure, $\angle P : \angle Q : \angle R = 1 : 2 : 3$. Find the value of $p : q : r$.

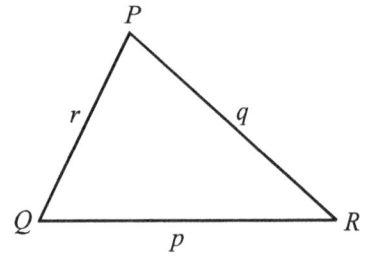

True or _False_

A circle is an *n*-sided polygon as *n* tends to infinity.

unit
9

The Beauty of Pi

π = 3? (1 Kings 7:23, 2 Chronicles 4:2)

If π = 3, this proves that the Bible is false!

If π equals 3, then scientists and mathematicians are lying to us!

In the Old Testament, in 1 Kings 7:23, or 2 Chronicles 4:2, the measurements about the altar built inside the temple of King Solomon suggests that π = 3!

Also he made a molten sea of ten cubits from brim to brim, round in compass, and five cubits the height thereof; and a line of thirty cubits did compass it round about. (1 Kings 7:23)

The above verse seems to suggest that

$$\frac{\text{Circumference}}{\text{Diameter}} = \frac{30 \text{ cubits}}{10 \text{ cubits}} = 3$$

So, π = 3!

Can the biblical π be reconciled with the mathematical pi?

Two conspiracy theories on the value of π are:

1. Mathematicians who have written textbooks are bound to lose money if they are proven wrong.

2. Mathematicians and scientists are keeping the "real value of pi" from us—perhaps because, like the Freemasons, they feel the need to keep the "truth" for themselves.

What's the value of pi on the moon?

Is Pi Rational or Irrational?

In most examination questions, it is not uncommon to ask students to take an approximate value of pi to be 3.14 or $\frac{22}{7}$, implying that π is an irrational number with no repeating pattern (3.1415...).

Yet, from the definition of pi, which is the ratio of the circumference of a circle to its diameter ($\pi = \frac{C}{D}$), this means that π is a *rational* number.

So, is π a rational or irrational number?

Practice

1. The length and width of a rectangle are x cm and y cm, respectively, where x and y are integers. If $xy + y = y^2 + 13$, find the maximum area of the rectangle.

2. What is the maximum number of squares in the picture?

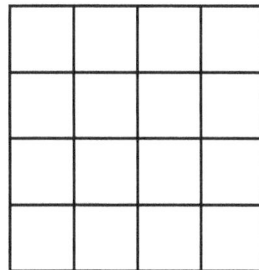

3. The perimeter of a rectangle is $16\sqrt{2}$ cm. Find the smallest possible value of the diagonal.

4. In the figure, $XM = MY = MZ = 5$, and $YZ = 6$.

   Find the area of the triangle XYZ.

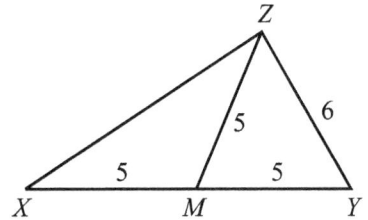

5. A quadrilateral has sides 1997 cm, 1998 cm, 1999 cm, and d cm. If d is a whole number, what is the largest possible value of d?

6. Job has 80 m of fence to enclose a rectangular dog pen. If the largest possible area is to be built for the dog to run against an existing wall,

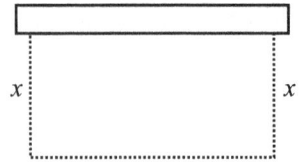

   (a) write an equation relating the area, A, and the width, x,

   (b) what width provides the largest possible area?

7. The faces of a cuboid have area 48 cm², 54 cm², and 72 cm².

    What is the length of the longest edge?

8. The perimeter of a right triangle is 180 cm and the length of the perpendicular height to the hypotenuse is 36 cm. Find the length of the hypotenuse.

9. In $\triangle PQR$, $PQ = 16$, $QR = 17$, and $RP = 18$. M is the midpoint of PQ and N is the perpendicular from P to QR. Find MN.

10. The figure shows two identical isosceles right triangle ABC.

    Squares $BDEF$ and $PQRS$ are inscribed in two ways, as shown. What is the ratio of the area of square $PQRS$ to the ratio of the area of square $BDEF$?

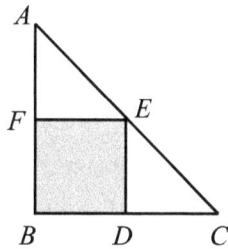

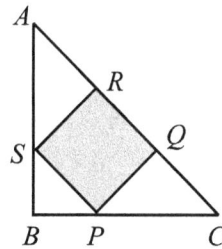

unit 10

$\diamond$

The Zero Option

The *Zero Option* is a technique popularized by the puzzlist Erwin Brecher to solve a special type of geometric problems, whose solution conventionally involves a tedious process, but which with proper insight may be solved relatively quickly. These problems generally appear to contain insufficient data for their solution; in fact, their solutions are independent of the missing information. On this assumption, the relevant dimension can be taken as zero, thus transforming the problem from a tedious exercise to a geometrical quickie.

A *geometrical quickie* may be defined as a geometrical problem which may be solved by laborious methods, but which with the right insight may be disposed of quickly. The zero option is somewhat similar to the *method of exhaustion* conceived by Eudoxus (400-437 BC) and extensively used by Archimedes (c. 287-212 BC), to determine the areas and volumes of geometrical figures, and to calculate the value of π.

Elementary Theorems in Plane Geometry

The zero option is not only favorable to geometrical quickies, but it also lends itself to proving elementary theorems in plane geometry. For a start, let's zero in on some geometrical quickies which elegantly make use of the zero option principle.

Worked Example 1

In the figure, given that $XY = 14$ cm, find the area of the shaded region, expressing your answer in terms of π.

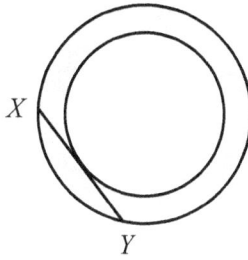

Solution

Since only the length of the chord XY is given, this problem appears to have insufficient information. The final result seems to be independent of the radii of the circles—this fact may be checked by using specific areas of knowledge (e.g., the Pythagorean theorem, in this case)—thus, suggesting the possibility of the use of the zero option.

Imagine the two circles decreasing in sizes, with the chord remaining 14 cm long, until the inner circle becomes a point circle. Its radius would then become zero, and the circle would simply be the point that is the center of the larger circle.

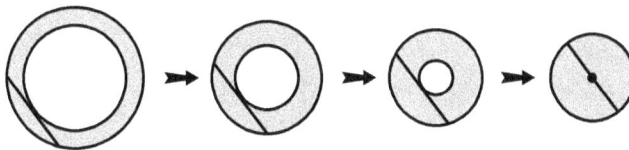

As the two circles decrease in size the chord becomes the radius of the larger circle.

The chord XY has now become the diameter of the larger circle, and its area is now 49π cm^2. This must be the required area between the two circles regardless of their sizes, as long as the chord is 14 cm long. Notice that the area of the annulus remains unchanged by the limiting process, using the invariant identity $R^2 - r^2 = 7^2$ (by Pythagorean theorem).

Compare this intuitive elegant solution that uses the zero option with the conventional solution that uses the Pythagorean theorem.

Let's look at another geometrical quickie that powerfully taps on the zero option for its solution.

Worked Example 2

A cylindrical hole of 12 cm long is drilled through the center of a solid sphere.

Find the volume remaining in the sphere.

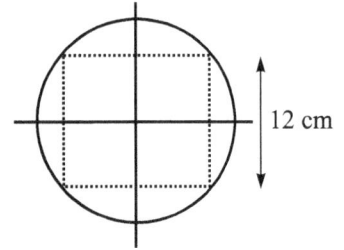

12 cm

Solution

If we let the diameter of the cylinder to be zero, then what remains in the sphere is the whole sphere.

Therefore, the volume remaining in the sphere is $\frac{4}{3}\pi R^3 = \frac{4}{3}\pi(6)^3 = 288\pi$ cm^3.

In other words, the residue is independent of the diameter of the hole or the size of the sphere.

Again, compare this ingenious zero option approach with its tedious algebraic counterpart, which requires the formula for the volume of a spherical cap.

The formula for the volume of a spherical cap is $V = \frac{1}{6}\pi h(3r_1^2 + h^2)$.

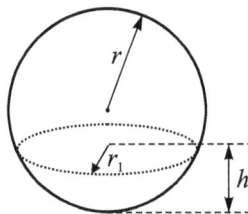

Practice

Use the zero option technique to solve the following questions, where relevant.

0. Two concentric circles are 10 cm apart. What is the difference between the circumferences of the circles?

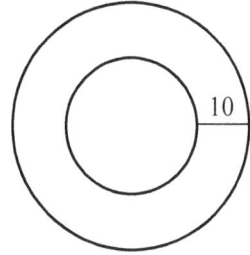

1. Using the fact that the formula for the circumference of a circle is $2\pi r$, derive the formula for the area of a circle, using the zero option technique.

2. The figure shows a rectangle *PQRS* inscribed in a quadrant of a square. Using the zero option, find the length of the diagonal *PR*.

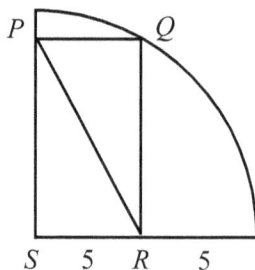

3. Show that when parallel lines are cut by a transversal, alternate interior angles are congruent (equal in size).

4. The figure shows two tangents to the circle drawn from point C. The lines AC and BC are equal, each of length 5 cm. Line DE is drawn, tangent to the circle at point P. Find the perimeter of the triangle CDE.

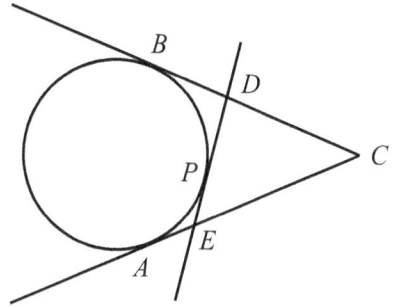

5. Using the fact that the formula for the volume of a cone is $\frac{1}{3}\pi r^2 h$, and the formula for the surface area of a sphere is $4\pi r^2$, derive a formula for the volume of a sphere.

6. A group of campers set out at 6:00 a.m. and return on the same track to their starting point at noon. If they average 4 km/h on level ground, 3 km/h uphill, and 6 km/h downhill, how far do they walk?

7. Two circles are concentric. A tangent to the inner circle forms a chord 12 cm long in the larger circle. Find the area of the ring between the circles.

8. A cylinder hole is drilled through a large, solid sphere of gold, producing a ring. The length of that cylindrical line is 6 cm. How much gold is left in the ring?

    (Volume of sphere = $\dfrac{\pi D^3}{6}$, where D is the diameter of the sphere.)

9. A long rope is tied along the equator of the Earth. Lengthen the rope by 1 meter that it is no longer tied around the Earth. If the rope is lifted equally around the equator so that it is uniformly spaced above the equator, will a mouse fit beneath the rope?

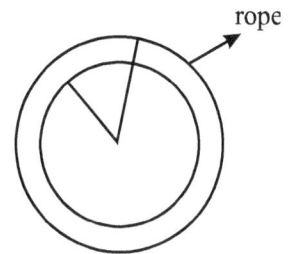

rope

10. A long wire is laid around the surface of the earth at the equator, and a similar wire is laid on the moon's surface around its equator. If each of these wires is raised 2 meters above the surface of the earth and moon respectively, supported by poles, which planet would need more wiring to implement this change?

God is always doing geometry.

— Plato (429–347 BC),
Greek philosopher

unit
11

────────────────────────────────────

The Golden Ratio
by Paper Folding

Take a strip of paper, say, about 1 or 2 cm wide, and make a knot. Then carefully flatten the knot, as shown below.

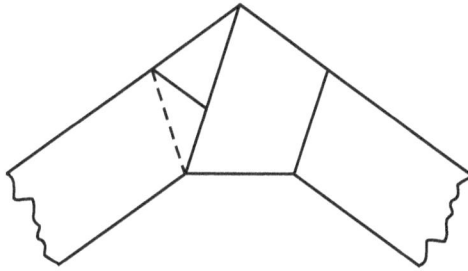

The resulting shape is a regular pentagon—one with its interior angles all equal and its sides all of the same length.

Now, imagine the pentagon with its diagonals. Can you show why they intersect each other in the golden ratio? In other words, show that point *D* divides line segment *AC* into the golden ratio, i.e., $\dfrac{DC}{AD} = \dfrac{AD}{AC}$.

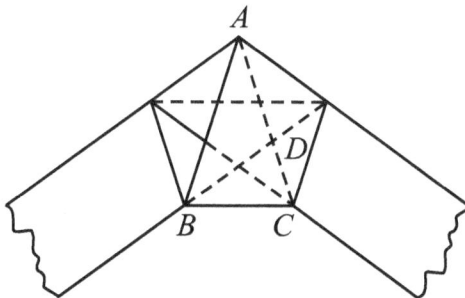

────────────────────────────────────

81

Theorem: The angle sum of a star (or pentagram) is 180°.

Here is a quick-and-dirty proof of the above theorem.

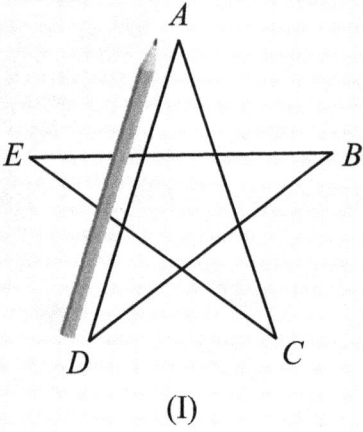

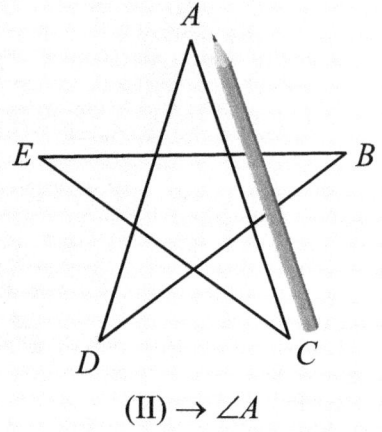

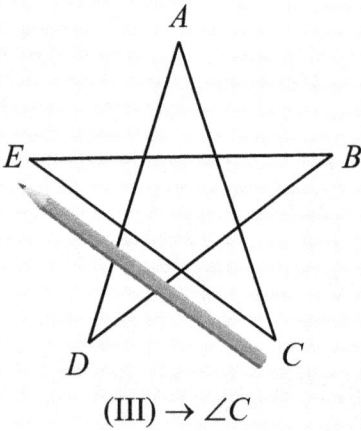

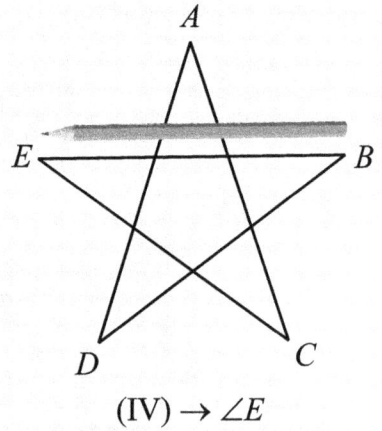

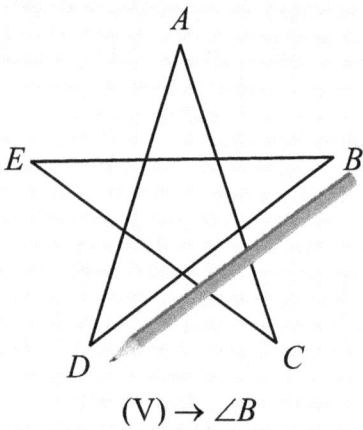

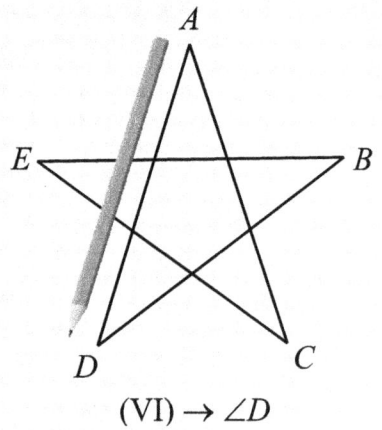

(I)

(II) → ∠A

(III) → ∠C

(IV) → ∠E

(V) → ∠B

(VI) → ∠D

(I) & (II): Place a pencil on line segment *AD* in the direction pointing at *A* and rotate it through ∠*A*, so that it is now on *AC* pointing at *A*.

(III): Then rotate it through ∠*C* so that it is now on *CE* pointing at *E*.

(IV): Rotate again through ∠*E* so that it is now on *BE* pointing at *E*.

(V): Again, rotate it through ∠*B* so that it is now on *BD* pointing at *D*.

(VI): Last, rotate it through ∠*D* so that it is now on *AD* pointing at *D*, which is the opposite direction of its starting position.

Therefore, the pencil has reversed its direction (*DA* to *AD*), which means it has rotated through an angle of 180º.

Hence, the angle sum of the star (or pentagram) which is the angle through which the pencil was rotated, angle by angle, is 180º.

Use a similar intuitive method to show that the angle sum of a triangle is 180º.

Practice

1. Given a 3 × 3 × 3-cm opaque cube divided into 27 one-centimeter cubes. What is the most number of 1-cm cubes that can be seen by someone from any point in space?

2. How many different squares can be formed by four dots that mark their areas?

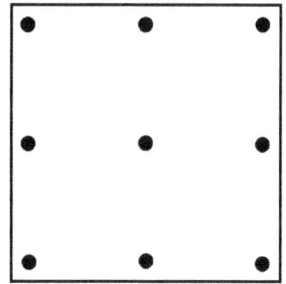

3. Two circles of radii 3 cm and 4 cm meet at right angles. If the overlapping regions are removed, find the difference between the two remaining areas.

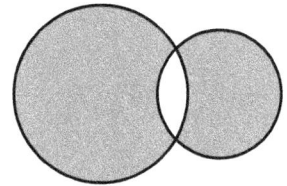

4. Two squares of sides 4 cm and 3 cm are cut into five pieces which are then rearranged into a larger square. What is the perimeter of the larger square?

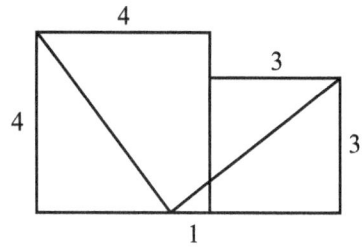

5. In the figure, $\frac{1}{5}$ of the small square and $\frac{2}{11}$ of the larger square are shaded. What is the ratio of the shaded area to the non-shaded area of the entire figure?

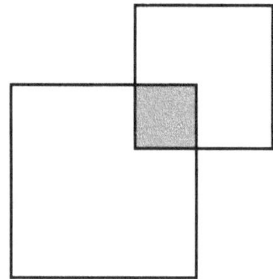

6. Ten points are marked around the circumference of a circle. What is the most number of chords that can be drawn joining these points such that no two chords cross each other?

7. In the figure, $\frac{3}{5}$ of the smaller circle and $\frac{1}{6}$ of the larger circle are shaded.

What fraction of the figure is shaded?

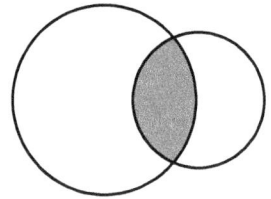

8. In the figure on the right, the radius of the big circle is three times the radius of the small circle. Assume that the "eyelids" are formed by two separate arcs of an even larger circle, what is the radius of that third circle relative to the "pupil"?

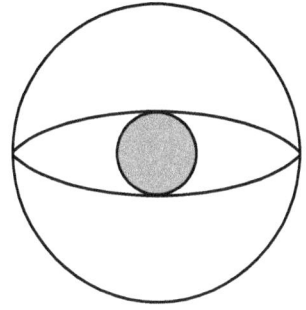

9. In the diagram, a circle is inscribed in an isosceles trapezoid, whose parallel sides have lengths 8 cm and 18 cm. What is the radius of the circle?

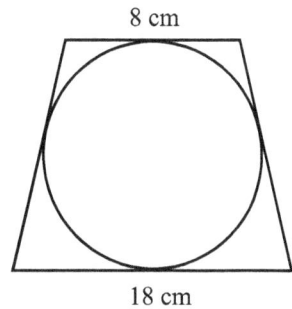

8 cm

18 cm

10. A farmer left a square plot of land to his three sons. The side of the land is expressed as a whole number of kilometers.

To the first son, he gave a square plot in the top-right corner, the size of which is again expressed in a whole number.

To the second son, he gave a plot that was square minus the top-right corner the first son owned. Again, the dimensions of the plot were in whole numbers.

The third child received the remaining plot whose area is equal to the area of the second son's plot.

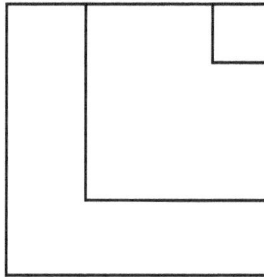

If the first son received the smallest piece possible plot of land, what are the dimensions of all three plots of land?

unit
12

The Ubiquity of Phi

The golden ratio with value $\frac{1+\sqrt{5}}{2}$ (≈ 1.618), symbolized by ϕ (pronounced "phi"), turns up in many arithmetical and geometric situations.

In the three examples below, show that the ratio of AC to AB is constant. All triangles are equilateral.

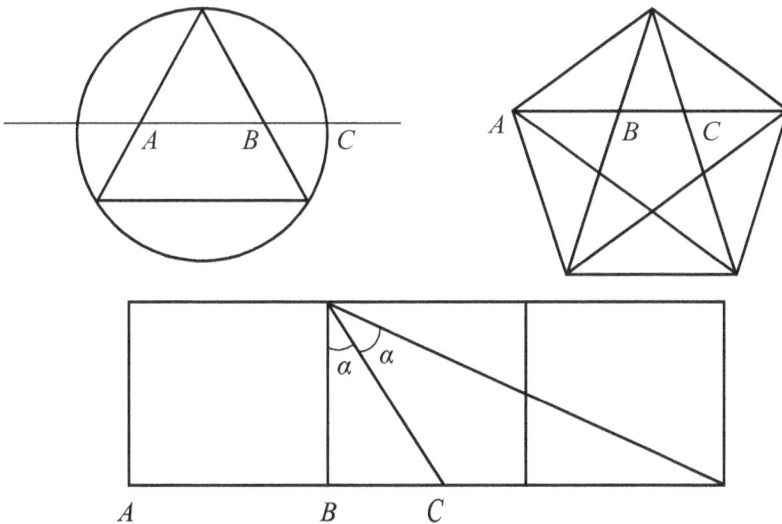

ϕ and $\frac{1}{\phi}$ are the roots of the quadratic $x^2 - x - 1 = 0$, which may be derived from the following proportion:

How would you divide a line segment into two parts such that the ratio of the larger part (x) to the shorter part is equal to the ratio of the length of the line segment to the larger part?

Practice

The following questions are all common in the sense that their answers yield the golden ratio or its reciprocal.

1. An isosceles triangle is inscribed in a square of side 2 cm. What is the radius of its incircle?

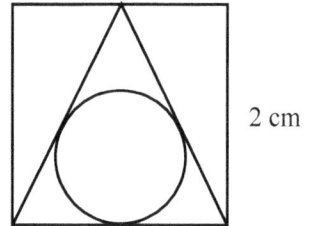

2 cm

---

2. In the figure below, an equilateral triangle is attached to a square of length one unit. Show that the ratio of *AC* to *AB* is the golden ratio.

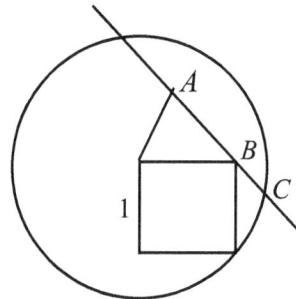

A

B

C

1

3. The figure shows an equilateral triangle *ABC* of side 2 units inscribed in a circle. Through the midpoints of two of its points, a line segment intersects the circle at points *M* and *N*.

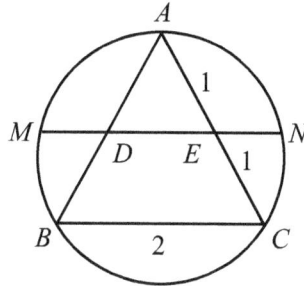

(a) Show that the line joining the midpoints of the two sides of the triangle is half the third side.

(b) Show that the ratio of the length of *MD* (or *EN*) to *DE* is $\dfrac{\sqrt{5}-1}{2}$.

4. Show that the golden ratio links these two infinite series.

$$1+\cfrac{1}{1+\cfrac{1}{1+\cfrac{1}{1+\dots}}} \qquad\qquad \sqrt{1+\sqrt{1+\sqrt{1+\sqrt{1+\dots}}}}$$

5. Show that the ratio of the radius of a circle to the side of an inscribed regular (10-sided) decagon is the golden ratio.

The Golden Ratio

A ratio, commonly known as *phi*, deified by recreational mathematicians, but denounced as devilish by theomaticians and biblical numerologists.

The question is: "Is Phi *Divine* or *Demonic*"?

6. Where does the parabola $y = x^2 - 1$ cut the straight line $y = x$?

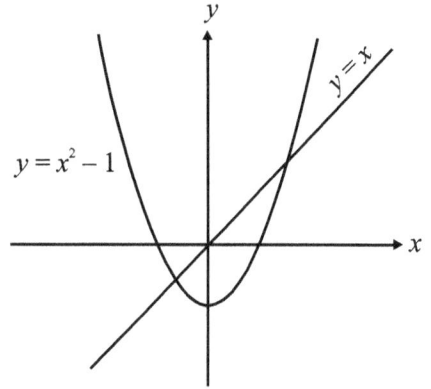

7. Where do the circle $x^2 + y^2 = 3$ and the hyperbola $xy = 1$ intersect?

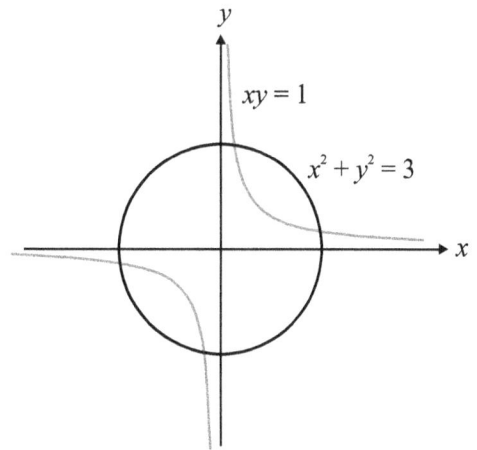

8.   The Pentagon and the Pentagram

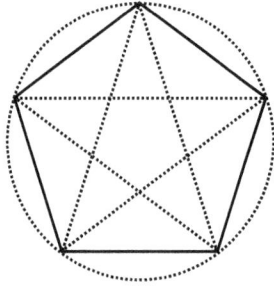

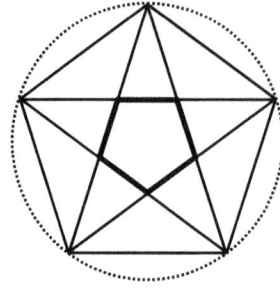

A *golden triangle* has the ratio of its side to its base equal to the golden ratio $\frac{1+\sqrt{5}}{2}$.

The Pythagorean Hippasus (c. 450 B.C.) showed that the ratio of a diagonal of the regular pentagon (i.e., a side of the regular pentagram) to the side of the pentagon is the golden ratio, ϕ (an irrational number $\frac{\sqrt{5}+1}{2}$). Verify that Hippasus was correct.

9. *ABCDE* is a pentagon inscribed in a circle with *AC* parallel to *ED* and $\angle ABE = 20°$. If *AD* and *CE* cross at *X*, what is the measure of angle *AXC*?

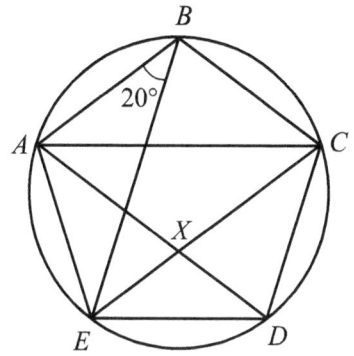

10. The pentagram (five-corner star) inscribed in a regular pentagon abounds with the golden ratio, ϕ, since it is made up of many golden triangles. In the figure, the regular pentagon has side length 1 unit and its diagonal has length $\dfrac{1+\sqrt{5}}{2}$ units, denoted by ϕ.

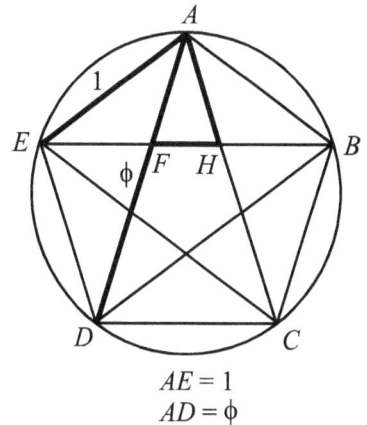

    (a) Show that $AH = \dfrac{1}{\phi}$ and $FH = \dfrac{1}{\phi^2}$.

    $AE = 1$
    $AD = \phi$

(b) Show that the ratio of the area of the larger pentagon $ABCDE$ to the area of the smaller pentagon is $\phi^4 : 1$.

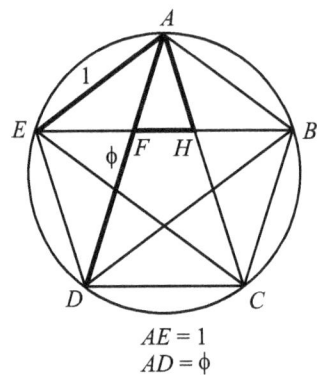

$AE = 1$
$AD = \phi$

(c) Show that the area of the larger pentagon $ABCDE$ to the area of the pentagram is $\dfrac{\phi^3}{2}$.

unit 13

Matchstick Mathematics

A recreational pastime for long-term convicts, matchstick puzzles have given them lots of challenge and fun during their solitary isolation. Indeed, there is more to matches than just lighting cigarettes.

Worked Example 1

In the figure below,

(a) remove 4 matches to leave 2 equilateral triangles,

(b) remove 3 matches to leave 2 equilateral triangles,

(c) remove 2 matches to leave 2 equilateral triangles.

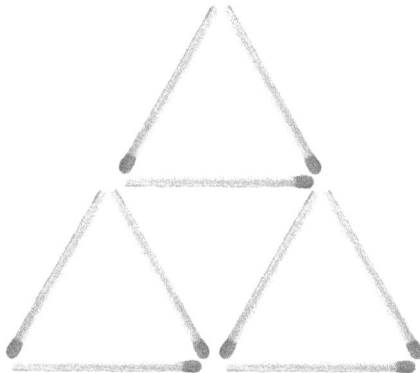

Solution

(a)

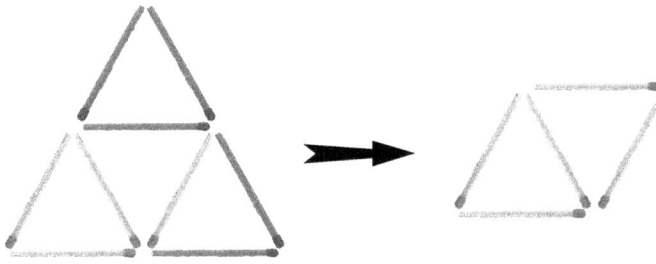

(b)

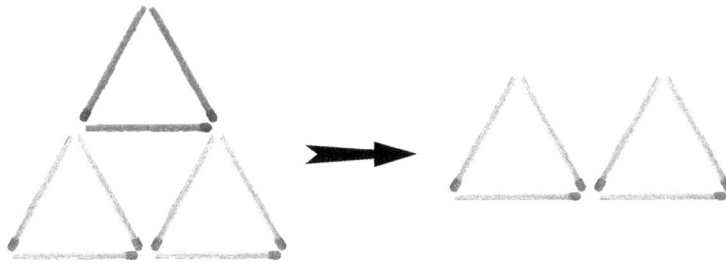

(c)

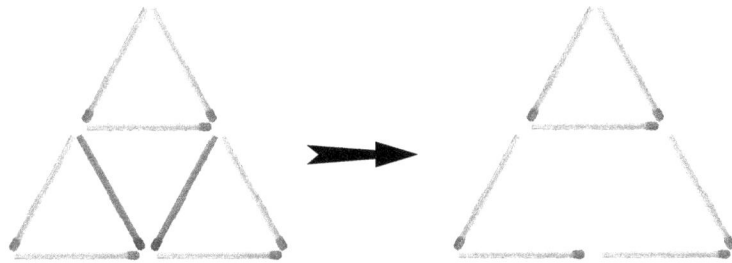

Worked Example 2

Make three identical squares from the pattern below by removing 3 sticks and moving 2 sticks.

Solution

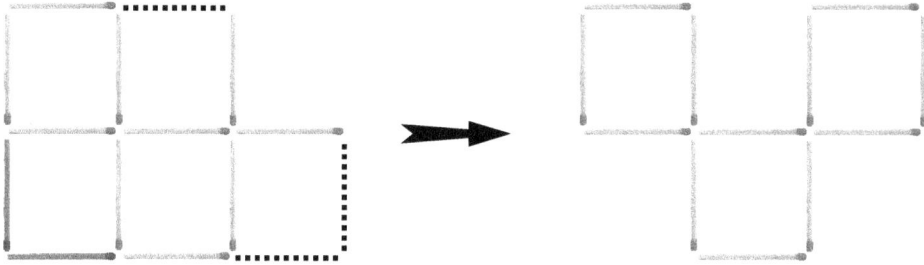

Practice

1. Move only 2 matches to change the three squares into four identical rectangles.

2. How would you move only one matchstick to change the house into two houses?

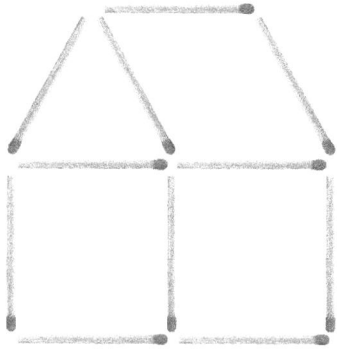

3. Move 3 matches to change the pattern given into a cube.

4. Move only 2 matches to change the pattern into four identical squares.

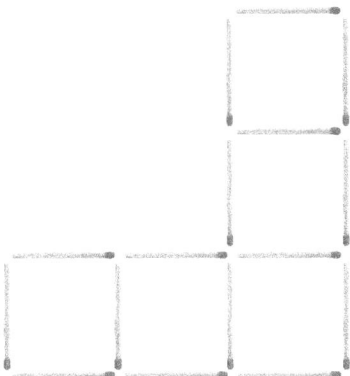

5. Make 3 identical squares from the 15 matches below by removing 3 sticks and moving 3 sticks.

6. In the pattern below, move 3 matches to make 3 squares.

7. In the pattern below, remove 2 matches, leaving only 2 squares.

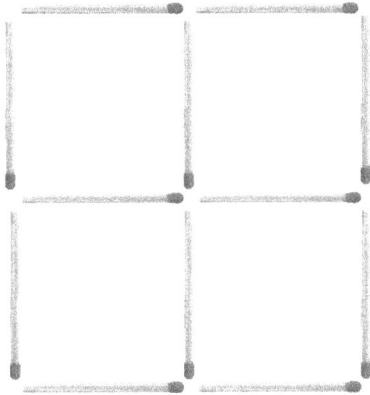

8. In the figure, 16 matches are used to form 5 identical squares. Move only 2 matches to turn the pattern into 4 identical squares.

9. In the pattern of 13 matches comprising of 6 identical triangles, remove 3 matches to leave 3 triangles.

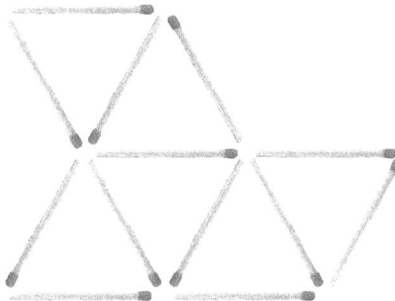

10. In the pattern below, move 4 matches to make 3 equilateral triangles.

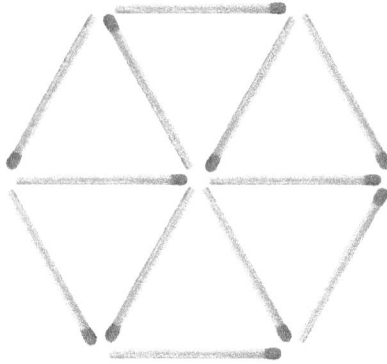

unit

14

The "Rolling Circle" Question

In the figure, the radius of circle A is three times the radius of circle B. Starting from the position shown, circle B rolls completely around the circumference of circle A which is stationary. How many revolutions will the center of circle B make before returning to its starting point?

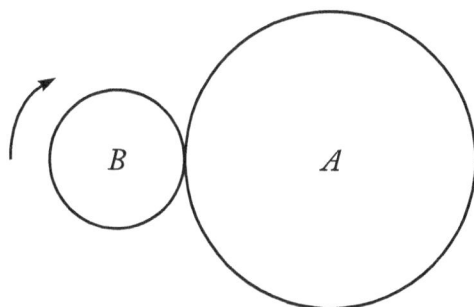

That circle B will make four revolutions on its axis as it traces a path around A before returning to its starting position is surprising even to the sophisticated problem-solver.

This little-circle-rolling-around-the-big-circle problem which appeared on a Scholastic Aptitude Test (SAT) test came into prominence, to the embarrassment of the designers of this item, that the correct answer was not included among its choices. The SAT choices were:

(A) $\dfrac{3}{2}$ (B) 3   (C) 6   (D) $\dfrac{9}{2}$ (E) 9

The SAT team's "correct" answer was (B).

This mathematical *faux-pas* forced the Educational Testing Service of Princeton to recalculate the scores of thousands of students on the mathematical section of the SAT.

This "Rolling Circle" question testifies the narrow mathematical view adopted by both students and experts. They deduced the intended method to be used and used it without noticing the common-sense complexities of the problem.

Understandably, the general argument to the rolling circle question is that since the circumference of *A* is three times that of *B*, *B* must make three revolutions about its own center. But if the experiment is carried out, it will be found that *B* makes four revolutions, as illustrated below:

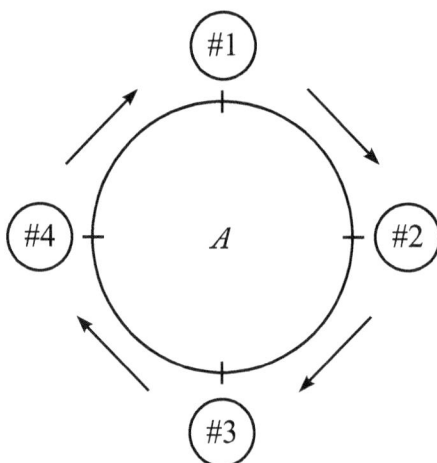

The number of revolutions would have been three if it were the center of *B* which had rolled around the circumference of *A*.

Since the circumference of *B* is one-third of *A*, this produces three rotations with respect to *A*, and the revolution adds a fourth revolution with respect to an observer from above.

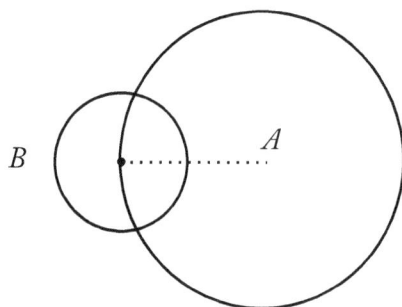

Note that the general formula for the number of rotations per revolution is $\dfrac{A}{B} + 1$.

In general, the number of rotations per revolution of two similar irregular objects is given by

$$\frac{\text{Perimeter of the path traced by the center of gravity of the rolling object}}{\text{Perimeter of the fixed object}}$$

The rolling circle problem is encountered in engineering, in the design of gear configurations.

Practice

1. Wheel A of diameter 1 cm rolls around a fixed wheel B of diameter 2 cm, as shown. How many revolutions about its axis will wheel A make in rolling once around wheel B?

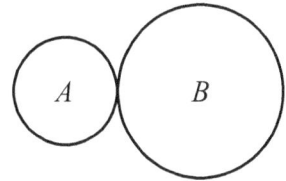

2. Two equal circular discs A and B are shown. If A is kept fixed and B is rolled round A without slipping, how many revolutions will B have made about its own center when it returns to its original position?

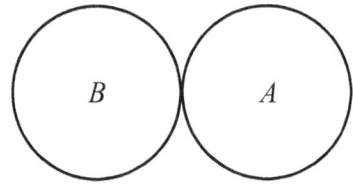

3. Wheel A has a diameter twice that of the fixed wheel B. If wheel A rolls around wheel B as shown, how many revolutions about its axis will wheel A make in rolling once around wheel B?

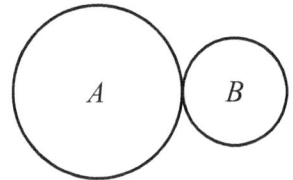

4. In the figure, the radius of circle A is three times the radius of circle B. Starting from the position shown, circle B rolls completely around the inside of the circumference of circle A, which is stationary.

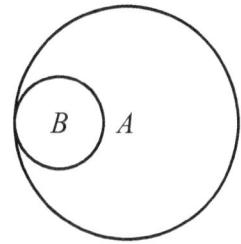

    How many revolutions will the center of circle B make before returning to its starting point?

5. The diagram on the right shows two spheres on either side of a hollow horizontal cylinder.

    The radius of the cylinder is three times the radius of either sphere. Both spheres are pushed around the cylinder until they return to their respective starting points.

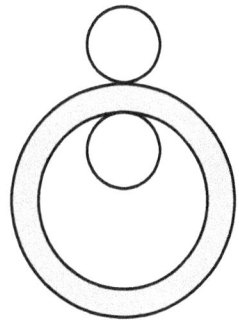

    How many more times will the outer sphere rotate as compared with the inner sphere turn through $360°$?

6. In the figure, circle B is half the radius of circle A.
    Circle B rolls around the fixed circle A in the plane.
    Fix a point on B.

    How does P move as it rolls around A?

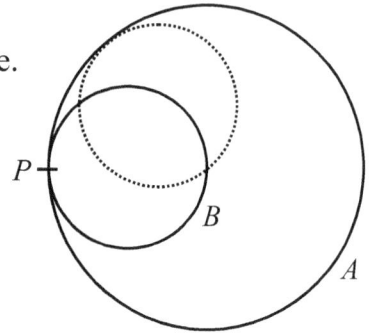

7. A cog-wheel of 8 teeth rotates on its axis
    round a fixed cog-wheel of 24 teeth.
    How many times does the small cog-wheel
    make around the big one?

8. The figure shows three congruent circles. How many circles will it take altogether to make a complete ring around the shaded circle?

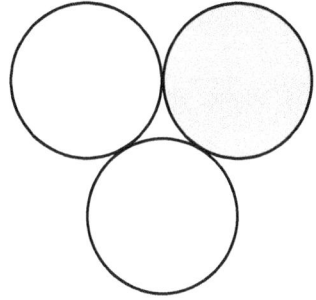

9. A circular disc of radius 2 cm is rolled without slipping around the outside of a triangle. The perimeter of the triangle is 12 cm. What is the length of the path traced out by the center of the disc?

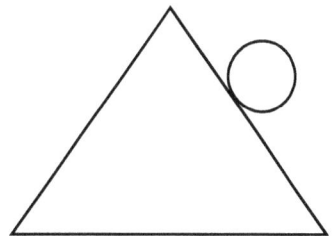

10. A circle of radius 1 cm is rolled without slipping around a square. The perimeter of the square is 20 cm. What is the length of the path traced out by the center of the circle?

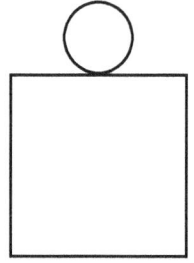

What is *wrong*?

surrounded on all sides

Square in shape

short in length

small in size

unit 15

Two Useful Circle Properties

(A) The angle formed by two chords intersecting within a circle is equal to one half the sum of the arcs intercepted by the angle and its vertical angle.

We need to show that $\angle ACE = \dfrac{1}{2}(\angle XOY + \angle AOE)$.

$$\angle ACE = \alpha + \beta \qquad \text{(ext. } \angle = \text{ sum of 2 int. } \angle\text{s)}$$

$$= \frac{1}{2}(2\alpha + 2\beta)$$

$$= \frac{1}{2}(\angle AOE + \angle XOY)$$

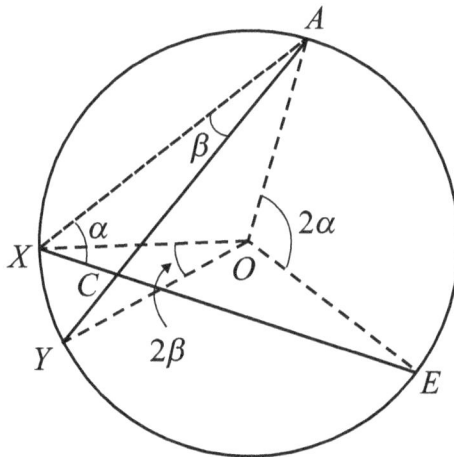

(B)   The angle formed by two lines intersecting outside a circle is equal to one half the difference of the arcs intercepted by the angle and its vertical angle.

We need to show that $\angle ACE = \dfrac{1}{2}(\angle AOE - \angle XOY)$.

$$\begin{aligned} \angle ACE &= \beta - \alpha \\ &= \frac{1}{2}(2\beta - 2\alpha) \\ &= \frac{1}{2}(\angle AOE - \angle XOY) \end{aligned}$$

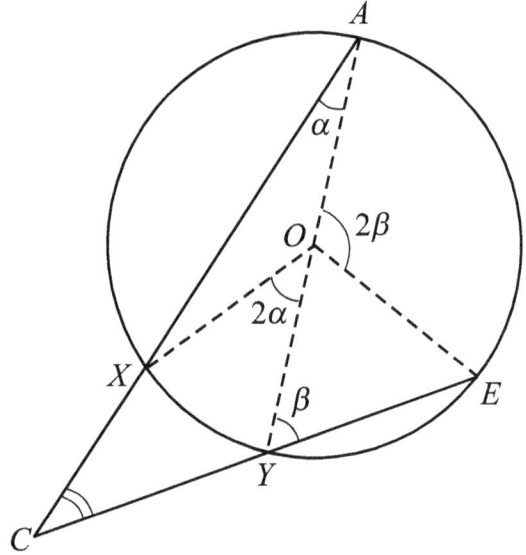

Practice

Use the circle properties above to solve questions 1–7.

1.   In the fi gure, *ABCDE* is a regular pentagon and *ACEBD* is a regular star. What is the sum of the interior angles of the star?

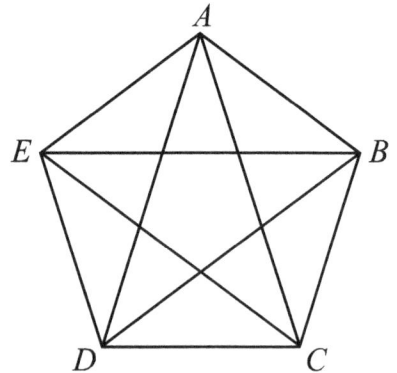

114

2. The figure shows a star inscribed in a circle. If the vertex C moves along the arc BD of that circle, a non-regular star is obtained. What is the sum of the internal angles of the star?

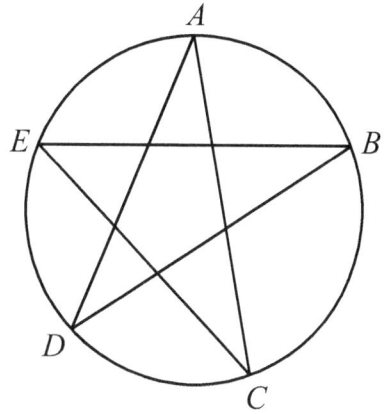

3. In the figure, $ACEBD$ is a non-regular star, with vertex C inside the circle. What is the sum of the interior angles of the star?

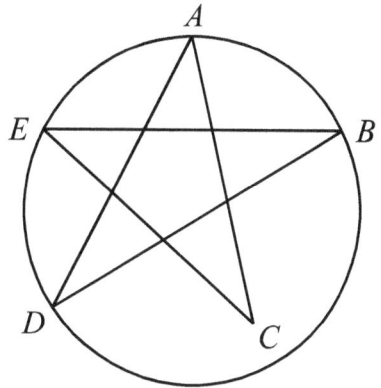

4. In the figure, point C of the star is outside the circle. What is the sum of the interior angles of the star?

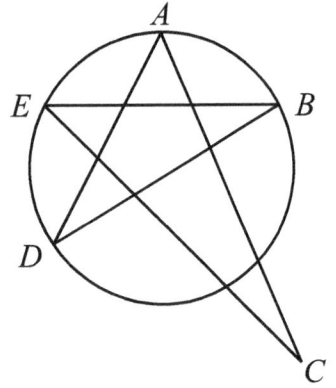

5. In the figure, points C and D of the star fall outside the circle. What is the sum of the interior angles of the star?

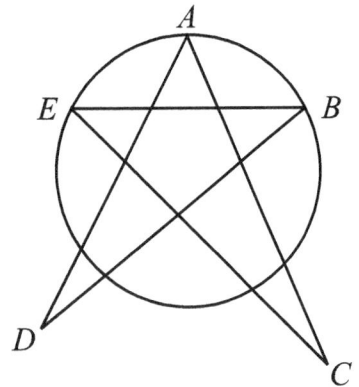

6. What is the sum of the internal angles of the star below?

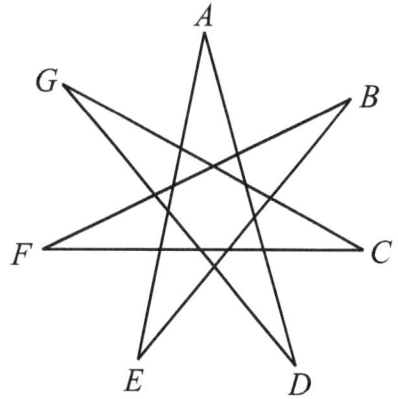

7. What is the sum of the internal angles of the star below?

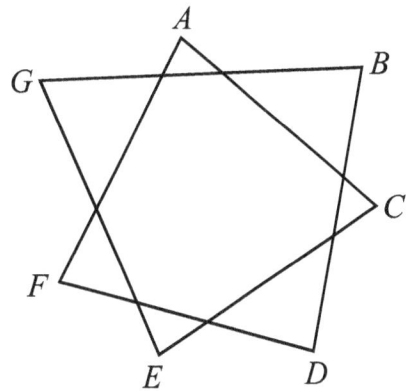

8. (a) Can one draw a six-pointed star in six straight line segments without lifting the pen?

    (b) Can one draw a five-pointed star by five straight line segments without lifting the pen?

9. Can one draw a seven-pointed star by seven straight line segments without lifting the pen?

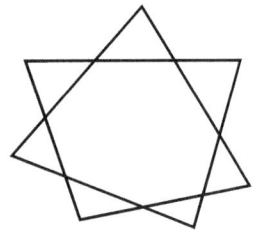

10. The figure shows a star made up of two overlapping triangles ABC and PQR. Two-thirds of triangle ABC is shaded. What is the ratio of the shaded area to the unshaded area of the figure?

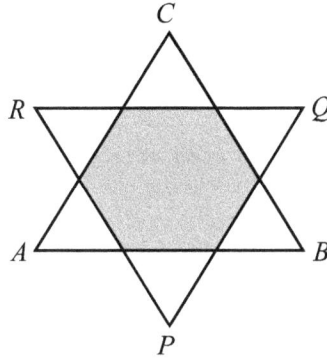

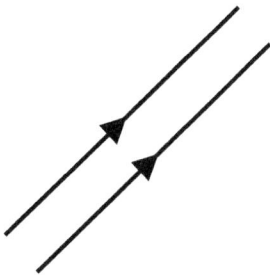

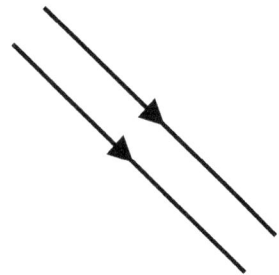

unit 16

Proving the Obvious

One common difficulty among novice geometry students is that many of the proofs seem unnecessary because the visual, itself, makes the result so obvious.

And, in geometry, not everything is "intuitively obvious" and there are geometric "facts" that not only seem to be wrong, but do not necessarily make sense without careful inspection.

How would you prove these "obvious results"?

1. **The perpendicular bisector of a line segment is equidistant from the endpoints of the segment.**

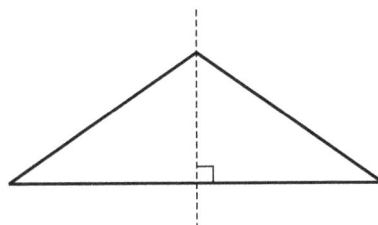

Is this just a semantic game? Are we merely playing with definitions?

2. **If a triangle is isosceles, then the angles opposite the equal sides are equal.**

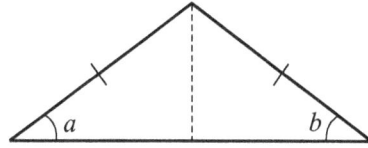

How can I know that two equal sides in a triangle always implies two equal angles?

3. **If a triangle is equilateral, then all three angles are equal.**

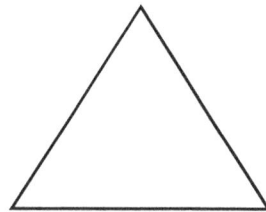

3 equal sides $\Rightarrow$ 3 equal angles?
or
3 equal angles $\Rightarrow$ 3 equal sides?

Faulty Intuition—A Counterintuitive Problem

Imagine that Earth, taken to be a perfect sphere with a radius r of 6378 km, is completely covered by a thin membrane. Now suppose that 1 square meter is added to the area of this membrane to form a larger sphere. By how much does the radius and the volume of this membrane increase?

Using the formulas for the volume of a sphere [$V = (4\pi r^2)/3$] and the area of a sphere ($A = 4\pi r^2$), respectively, it turns out that if the area of the cover is increased by 1 square meter, then the volume it contains is increased by about 3.25 million cubic meters. However, the new cover would not be very high above the surface of the planet—only about 6 nanometers!

How to Be "Geometrically Literate"

Visit second-hand bookshops, and look for old geometry textbooks.

Deductive geometry will give you a definite edge when it comes to many "recreational" mathematics problems. Besides, you often have a different and more elegant way of approaching mathematical problems in general.

Savor the rigor of geometry with fervor —that cold austere beauty.

Practice

1. In the figure, $AN = BM = AB$, $\angle C = 35°$. Find $\angle APB$.

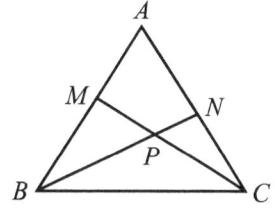

2. A square peg just fits in a round hole. What fraction of the hole is occupied by the peg?

3. In the diagram, the line divides the area of the rectangle in the ratio 1 : 4. What is the ratio of $a : b$?

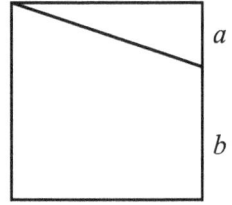

4. In the figure, the side of the square is 2 cm. What is the radius of the circle?

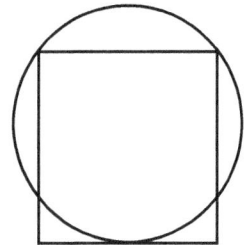

5. How many squares can you create in this figure by connecting any four dots?
   Note: The corners of the square must lie upon a grid dot.

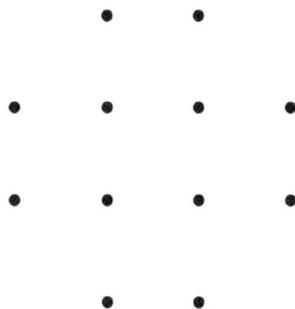

6. In the figure below, line segment AB is 1 cm long, is tangent to the inner of two concentric circles at A and meets the outer circle at B. What is the area of the annular region between the circles?

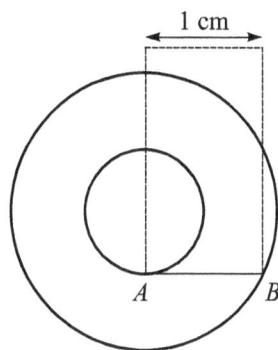

7. In the figure, $\triangle PQR$ is divided into 4 parts, with $QM = RM$.

   The ratio of the areas of the parts are such that $A : B = 3 : 2$ and $C : D = 7 : 5$. What is the ratio of area B to area D?

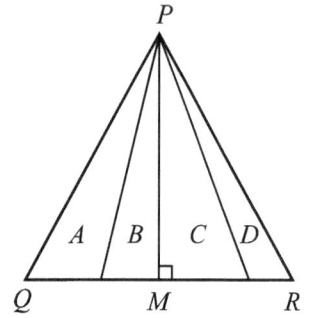

8. The figure shows a parallelogram of area 60 cm². A segment is drawn from one vertex to the midpoint of an opposite side. The diagonal is drawn between two other vertices. What are the areas of the four regions formed?

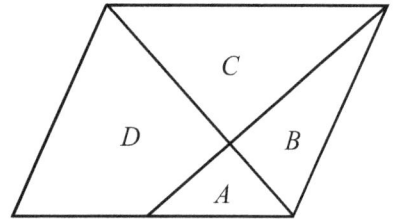

9. How many squares are there in the figure on the right?

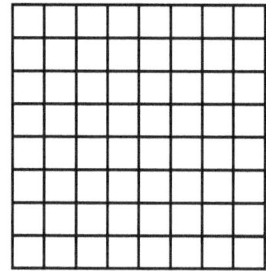

10. In the figure below, the two circles touch at A. The larger circle has its center at B. The width of the crescent between points C and D is 9 cm, and the distance between points E and F is 5 cm. What are the diameters of the two circles?

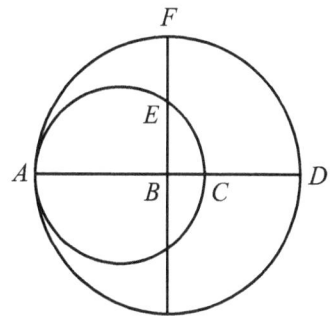

unit 17

SANGAKU—FUN WITH JAPANESE TEMPLE GEOMETRY

From 1639 to 1854, Japan lived in strict, self-imposed isolation from the West. During this period of national seclusion, a kind of native mathematics thrived. Devotees of math (samurai, merchants, and farmers) would solve a wide variety of geometry problems, inscribe their efforts in delicately colored wooden tablets, and hang the works under the roofs of shrines and temples.

The Japanese word, *wasan*, is used to refer to native Japanese mathematics, as compared to *yosan*, or Western mathematics. These *sangaku*, a word that literally means mathematical tablet, may have been acts of homage or a thanks to some guiding spirit, or simply challenges to other worshippers: *Solve this one if you can!*

Sangaku deal with ordinary Euclidean geometry but the problems differ from those that appear in a typical geometry textbook. Circles and ellipses play a prominent role in these sangaku problems: circles within circles, ellipses within circles, circles within triangles, spheres within pyramids, or ellipsoids surrounding spheres.

Who produced the sangaku? Were the theorems so beautifully drawn on wooden tablets the works of professional mathematicians or amateurs?

128

A few of these recreational geometry problems predate known Western results, such as the Malfatti theorem, the Casey theorem, and the Soddy hexlet theorem.

Let's look at some of these Japanese temple geometry problems. Appreciate their beauty and simplicity. Challenge yourself to solving them.

Practice

1. A circle is inscribed inside a 3-4-5 right triangle. What is the radius of the circle?

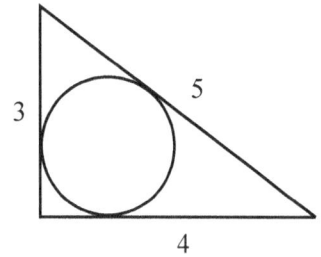

- -

2. The sides of the triangle are 3 m, 4 m, and 5 m. What big is the inscribed square?

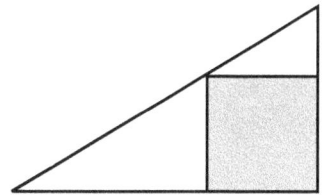

3. Three circles touch each other externally and have a common tangent, as shown. If the radii of the large circles are 4 cm and 1 cm respectively, what is the radius of the smallest circle?

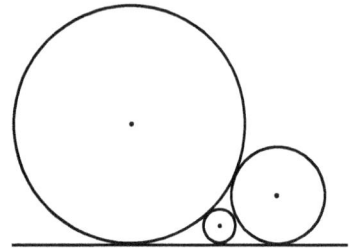

4. The figure contains two squares inside an isosceles triangle. When one of the squares becomes bigger, the other becomes smaller. Show that the sum of their sides ($a + b$) is a constant.

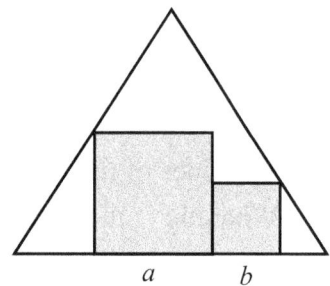

5. In the figure, two tangent semicircles are inside an isosceles triangle. The centers of the circles lie on the triangle's base. Show that the sum of their diameters $(d_1 + d_2)$ is a constant.

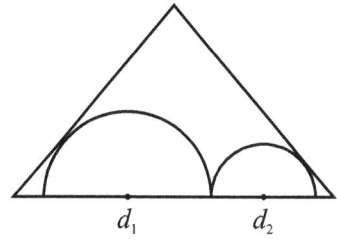

6. The figure shows two squares inside a semicircle. Show that the sum of the squares of their sides is constant.

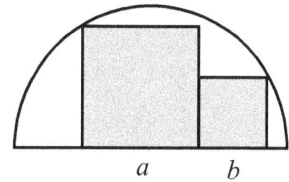

7. If the radius of the large semicircle is 6 cm, what are the radii of the other circles and semicircles?

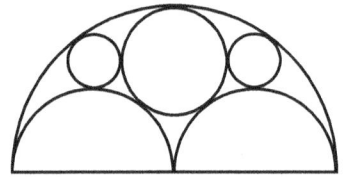

8. In the figure, which is bounded by a square, show that the two upper circles have the same radius as the three lower circles.

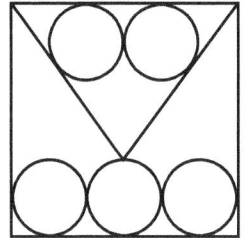

Many of the *Sangaku* tablets were lost through the destruction of some of the Japanese temples; today, there are about 900 of these tablets left.

9. The figure shows a semicircle with two smaller semicircles constructed along its diameter. A line l is drawn from the point on the diameter where the two circles meet and extended perpendicularly from the diameter to the circumference of the large semicircle. What is the shaded area?

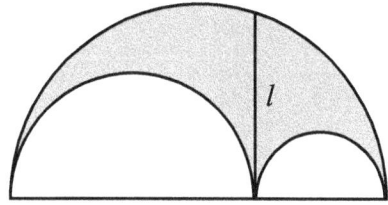

10. Two small semicircles are of equal size and their centers lie on the diameter of the big semicircle. The grey circles fit tightly. Show that the sum of the diameters of the grey circles is a constant.

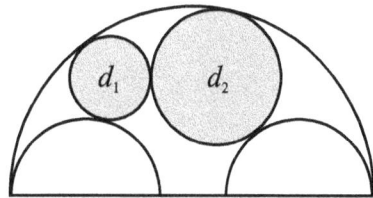

unit 18

THE PYTHAGOREAN THEOREM VISUALIZED

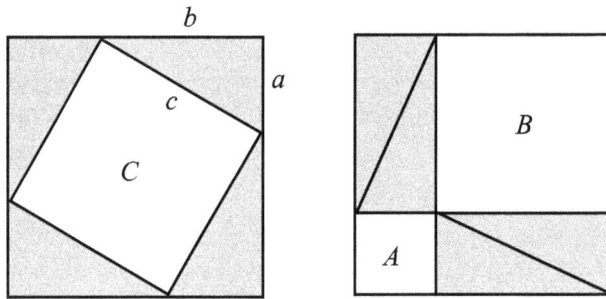

Square C is the square of the hypotenuse (c) of the triangle. Squares A and B are the squares of the legs (a and b) of the triangle.

Subtracting the four triangles from each square on the left and right, we are left with Square C on the left and Square ($A + B$) on the right. So they must be equal:

$$C = A + B$$

$$c^2 = a^2 + b^2$$

Does the theorem look obvious?

A visual representation makes the theorem more convincing—it gives the *why* of its existence!

Worked Example 1

In the figure, $PQ = QS = 5$ cm, $PQS = 90°$ and $SR = 2PS$. What is the length of QR?

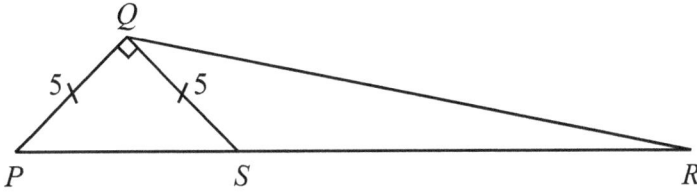

Solution

Let the perpendicular from Q meet PS at T.

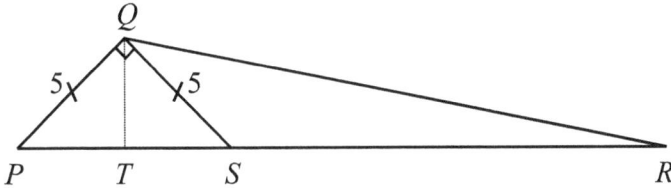

In $\triangle QTS$, $TSQ = 45°$ ($\triangle PQS$ is an isosceles right $\triangle$).

　　Thus $QT = TS$

In $\triangle QTS$, by the Pythagorean Theorem,

$5^2 = QS^2 = QT^2 + TS^2$

　　　　$= 2QT^2$　　　　　$(QT = TS)$

$QT^2 = \dfrac{25}{2}$

Now, $SR = 2PS$

　　　　$= 4TS$

　　　　$= 4QT$

Therefore, $TR = TS + SR$

$$= QT + 4QT$$
$$= 5QT$$

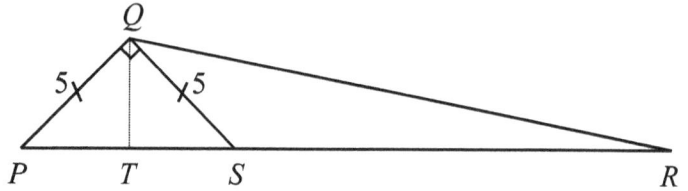

In $\triangle QTR$, by the Pythagorean Theorem,

$$QR^2 = QT^2 + TR^2$$
$$= QT^2 + 25QT^2$$
$$= 26QT^2$$
$$= 26 \times \frac{25}{2}$$
$$= 13 \times 25$$
$$QR = \sqrt{13 \times 25}$$
$$= 5\sqrt{13}$$

The length of QR is $5\sqrt{13}$ cm.

Worked Example 2

In the figure, $RS = 10$ cm, $ST = 6$ cm and $RT = 8$ cm. What is the area of rectangle $PQRS$?

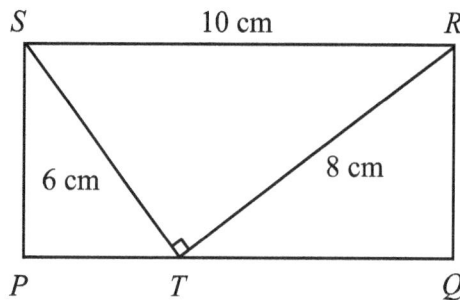

Solution

Since $6^2 + 8^2 = 10^2$, $\angle STR = 90°$

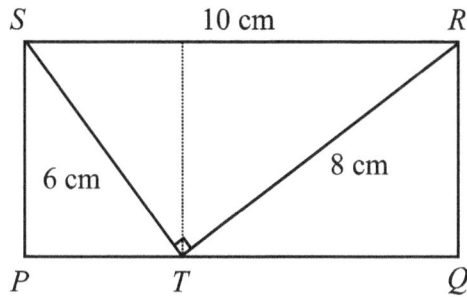

Area of rectangle *PQRS*

= 2 × area of triangle *RTS*

$= 2 \times \dfrac{1}{2} \times 6 \times 8$ cm^2

= 48 cm^2

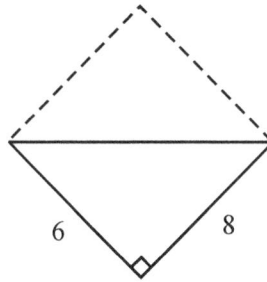

Practice

1. In a right triangle, the lengths of the adjacent sides are 550 units and 1320 units. What is the length of the hypotenuse?

2. In the figure, the right triangle *ABC* contains a square *AFDE* and two small right triangles *BED* and *DFC*. Given that *BD* = 5 cm and *DC* = 3 cm, what is the total area of triangles *BED* and *DFC*?

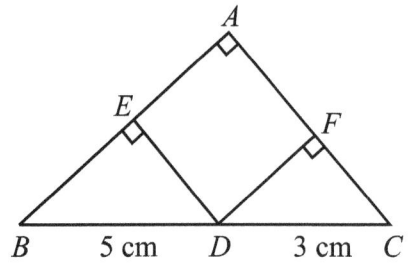

    (a) Without the Pythagorean Theorem, find the total area of triangles *BED* and *DFC*.

    (b) Show that the total area of triangles *BED* and *DFC* is $\dfrac{17}{15}$ the area of square *AFDE*. Hence, use the Pythagorean Theorem to verify your answer to (a).

3. In the figure, four isosceles right triangles are removed from the four corners of a square paper, leaving a rectangle. The total area of the cut-off pieces is 128 cm². What is the length of the diagonal of the rectangle?

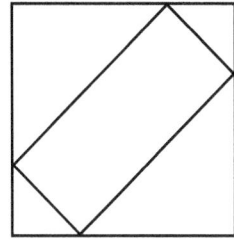

4. In the figure, $PM = MQ = MR = 5$ cm and $QR = 6$ cm. What is the area of triangle PQR?

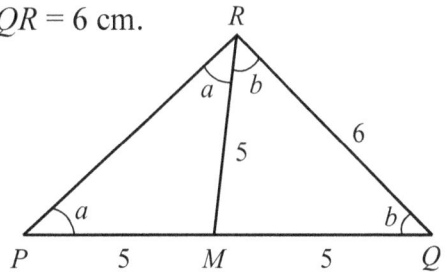

5. The three sides of a triangle are in whole numbers. Its perimeter is 8 cm. What is its area?

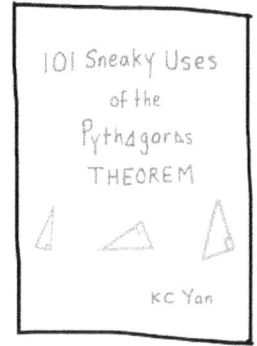

101 Sneaky Uses
of the
Pythagoras
THEOREM

KC Yan

---

6. The diagram shows two semicircles. *CD* is a tangent to the smaller semicircle and is parallel to *AB*. If *CD* is 12 cm, what is the area of the shaded region?

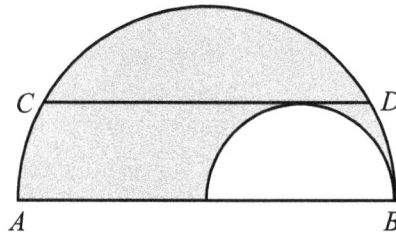

7. If triangle *ABC* is not a right triangle, what is its area?

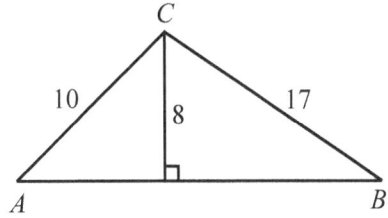

8. The advertisement for a 30-inch TV set refers to the length of the diagonal of the TV screen.

    (a) If the width of a TV screen is about 1.8 times its height, what are its dimensions?

    (b) How much larger (in terms in area) is a 30-inch set than a 20-inch set?

9. A banker decides to surprise his family members, by building a circular pool on a triangular plot of land, as shown. What is the largest possible radius of the pool that can fit in?

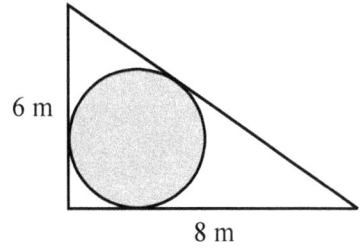

6 m

8 m

10. What is the shortest distance from P to QR in terms of q and r?

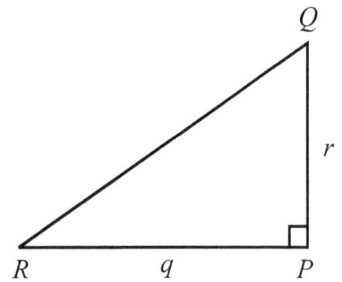

Q

r

R q P

unit 19

VISUALIZING INFINITY

Let's tap on the power of visualization to help us determine the values of some infinite series. These pictures serve as some "proofs without words," or what is commonly termed the "look-see proofs."

$$\frac{1}{3} + \frac{1}{9} + \frac{1}{27} + \frac{1}{81} + \ldots = ?$$

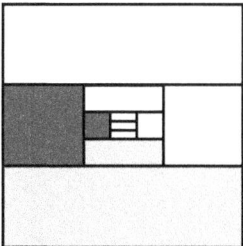

$$\frac{1}{4} + \frac{1}{16} + \frac{1}{64} + \ldots = ?$$

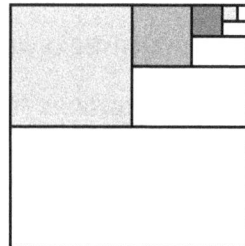

$$\frac{3}{4} + \frac{3}{16} + \frac{3}{64} + \ldots = ?$$

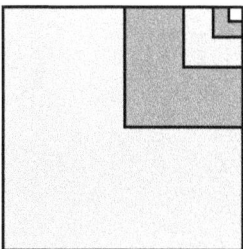

$$\frac{1}{4} + \left(\frac{1}{4}\right)^2 + \left(\frac{1}{4}\right)^3 + \ldots = ?$$

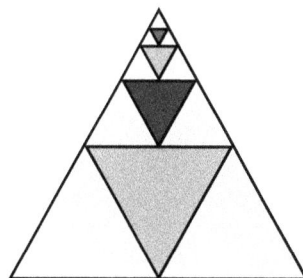

143

Practice

1. How many different ways can six arrows be put on the faces of a cube?

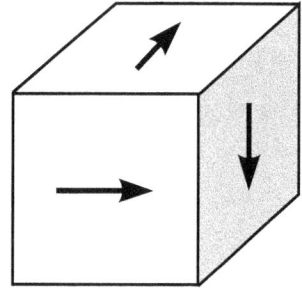

2. Twelve matchsticks are used to form four equal squares, as shown.

   Move and rearrange four matches to form three equal squares of the same size as the ones shown here.

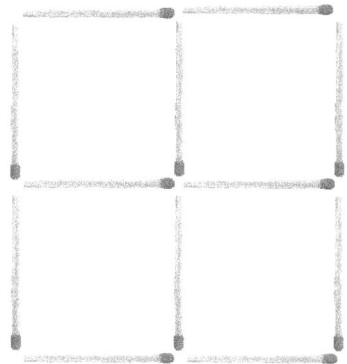

3. (a) How many squares of different sizes can be formed by connecting the dots that form this grid?

   (b) What is the total number of squares that can be created by connecting the pattern dots?

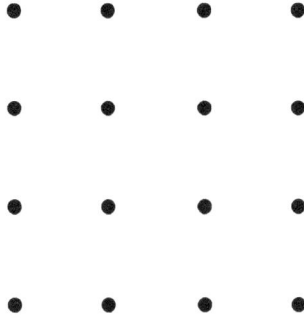

4. The figure shows a sequence of circles inside a circle, which goes on indefinitely. How is the area of the big circle of radius one unit related to the areas of the four sets shown within?

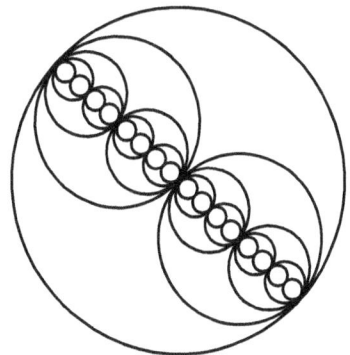

5. What do these figures have in common?

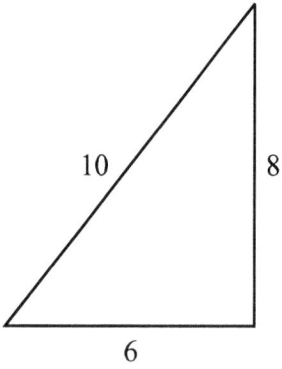

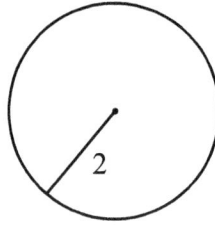

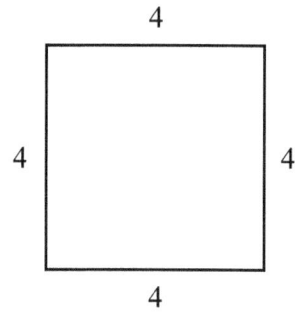

6. Why are manhole covers round?

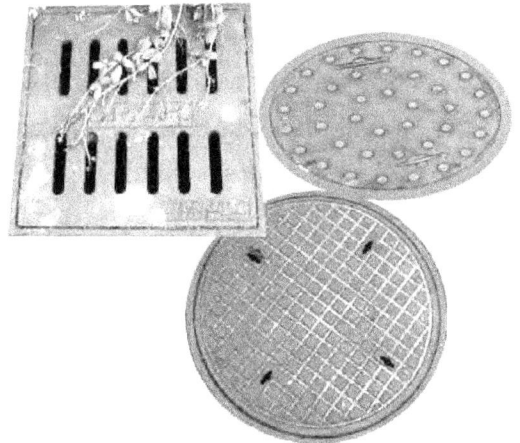

7. A square pyramid with each edge of length 1 cm is added to every face of a cube of side 1 cm to form a solid star. How many edges does the new solid have?

8. The net is folded to form a cube with the numbers on three faces meeting at a vertex being multiplied together. What is the largest such product for the vertices of this cube?

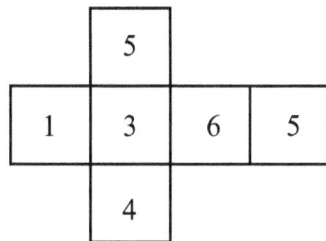

	5		
1	3	6	5
	4		

9. In the figure below, $AE = \frac{1}{3}AB$, $CD = \frac{1}{3}CA$ and $CG = GF = FB$.
   What fraction of the figure is unshaded?

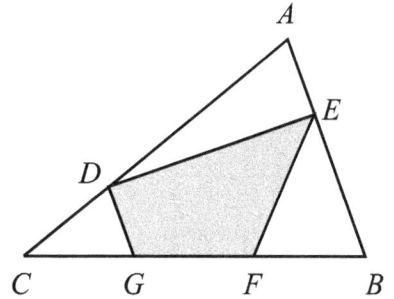

10. A bar is made up of 23 cubes fitted together. Where in the bar can you make just four "breaks" in order to obtain any number of cubes from 1 to 23 without having to perform any other additional "breaks"?

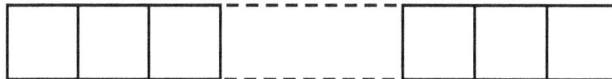

unit 20

12. There are no left triangles, only right triangles.

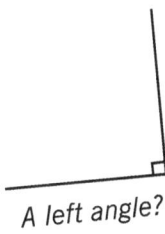

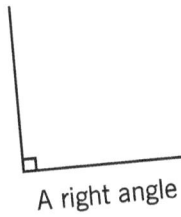

A left angle?   *A right angle*

11. An enlargement can be smaller.

10. A figure of order one has zero rotational symmetry.

0

Order = 1

Rotational symmetry = 0

9. **A straight line is a curve.**

Mathematicians have a strange definition for a curve.

8. **A line has no thickness.**

Ask Dr. Euclid why!

Can a rectangle be a square?

7. **A SQUARE IS A RECTANGLE.**

6. An equilateral triangle is an isosceles triangle.

5. A set square is triangular.

4. **Pythagoras may be a legend!**

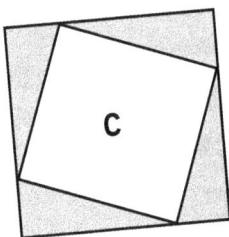

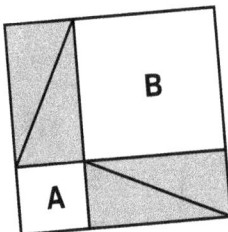

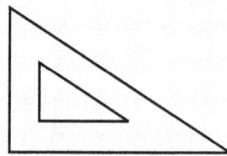

C

B

A

3. A triangle of circles is not a triangle.

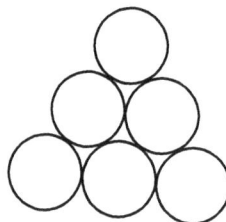

2. A rectangled square is NOT a rectangle.

1. **A squared rectangle is not a square.**

Practice

1. There are 9 dots in a square. Move only 4 of these dots to obtain a smaller square that has only 3 dots in each row and column.

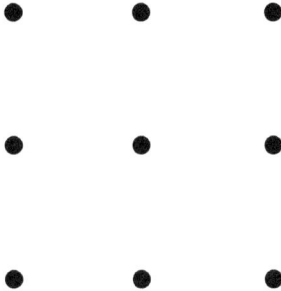

    •   •   •

    •   •   •

    •   •   •

2. Given three identical rectangular bricks and a ruler, without using any formula, how would you find the length of the brick s diagonal?

3.   (a)   What is so special with the right triangle below?

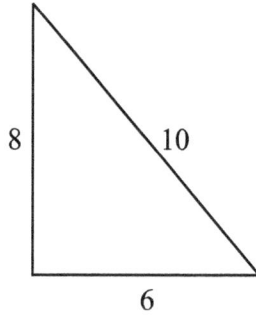

   (b)   Given that the sides are in whole numbers, find another right triangle having the same property.

4. Abel and Cain are inside a large field which has the shape of an equilateral triangle.

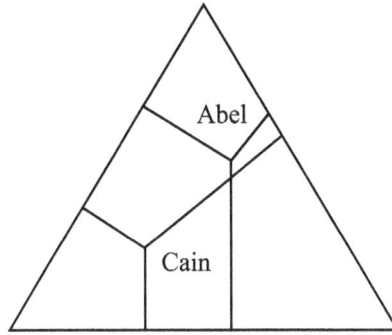

Abel says that they must stand in the middle of the field so that their total distance from all three sides is as small as possible.

Cain says that they can stand anywhere inside the triangle.

Who is right?

5. How many triangles of different sizes and shapes can be formed using the vertices of a cube?

6. How many perfect squares of all possible sizes are hidden in the shape of the dots?

   Note: The dots should be in the corners of the squares.

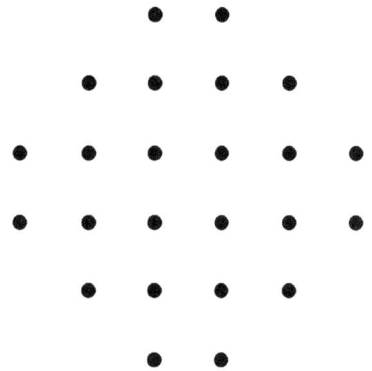

7. Find a non-square rectangle whose perimeter equals its area. All sides are expressed in whole numbers.

8. The figure shows a number of semicircles, with the horizontal diameter of the circle being cut into three equal lengths. If the area of the circle is 6 cm^2, what is the area of the shaded part?

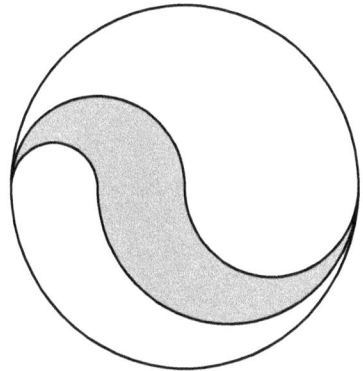

9. Draw the angle bisectors of any parallelogram.

    What shape do they make inside the parallelogram?

    Why does this always happen?

---

10. The marked angles are equal. Why must the two shaded areas be equal?

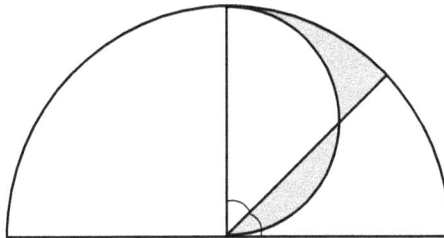

HINTS

Peek at them, unless you need to!

Hints for Units 1–20

Unit 1 (page 3)

1. Rotate the smaller triangle until its vertices touch the bigger triangle.

2. No Pythagorean theorem is required to solve this question.

3. Rearrange the shaded parts to form a cross.

4. What about rotating one square until two of its sides touching the center of the other square are at right angles to each other?

5. You needn't calculate the length and the width of rectangle x.

6. Draw horizontal and vertical dotted lines passing through the common vertex of all four triangles to find common areas of rectangles.

7. Try it out, with two cutouts: one circle whose radius is three times that of the other circle.

8. Stack the cut pieces on top of each other in subsequent cuttings.

9. Unless otherwise stated in the question, never assume that an angle that looks like a right angle to be one.

10.   Divide the figure into a right triangle and a rectangle.

Unit 2 (page 11)

1. *Method 1:* Circumscribe a larger square around the larger circle. Then rotate the smaller square such that its vertices touch the midpoints of the larger square. Finally, use the relationship: Area of smaller square : Area of larger square = Area of smaller circle : Area of larger circle.

Method 2: Use the Pythagorean theorem to compare the ratios of the two circles.

2. If x is the width between the two squares, show that $(17 + 2x)^2 - 17^2 = 111$.

3. Use the Pythagorean theorem to find the perpendicular height (or altitude) of the equilateral triangle.

4. Ratio of areas = (Ratio of lengths)^2.

5. Use the Pythagorean theorem, or the *Zero Option* discussed in unit 10, to solve this question.

Unit 2 (page 13)

6. What about rotating one of the squares?

7. You needn't find the area of each unshaded region and take the difference. Show that the answer is the difference between the areas of the two circles.

8. Label the smaller pentagon *ABCDE* (inside the star), with the lower vertex as *A*, in an clockwise direction. Then label the vertices of the larger pentagon *FGHIJ*, with the bottom right vertex as *F*, in an counterclockwise direction.

Considering each side of the larger pentagon's side as the base of a triangle, there would be 6 triangles. Now, for every pair of adjacent sides, there is a common triangle. So, if we consider the 6 triangles on one side, then three other sides will each have 5 triangles and the last side will have 4 triangles, giving a total of 25 triangles.

Next, consider the "star" inside the pentagon. Observe the 5 triangles: *ABJ, BCI, CDH, DEG, EAF*. There are also 5 other triangles (each point of the small pentagon in the star as a vertex): *AIG, BHF, CGJ, DIF, EJH*. How many triangles are there in total?

9. Draw two other quarter arcs above the existing ones inside the square, then move some shaded area around to find any relationship.

10. Draw dotted lines joining the vertices of the five-shaded figure.

Unit 3 (page 20)

1. Form a right triangle with one vertex touching the center of the larger circle and one vertex touching the intersecting corners. Then use the Pythagorean theorem.

2. Observe that half the surface area of the pipe has the same area as one and a half times the surface area of a cylinder of height 40 cm and diameter 20 cm.

3. Volume of fulcrum = Volume of cone with radius 3 cm – Volume of cone with radius 2 cm (that has been cut off).

5. (a) Draw a horizontal line segment across the figure.
   (b) Draw line segments from one vertex to another vertex.

6. Observe that the height of the column is equivalent to 5 complete spirals, and use the Pythagorean theorem to find the length of one spiral.

7. Draw a radius to each side of the triangle touching the circle.

8. In the hexagon, draw line segments joining opposite vertices; in the triangle, draw line segments joining the midpoints of the three sides.

Unit 3 (page 24)

9. Look for two whole numbers whose sum is 20, and whose product is 96—you may use *guess and check* to find them.

10.   You needn"t calculate the dimensions of the length, width, and height to find the volume of the cuboid.

Unit 4 (page 26)

1. *As simple as 1, 2, 3!*

2. If x is the radius of the smallest circle, show that the radii of the other two circles are $(6 - x)$ and $(8 - x)$ respectively.

3. Observe that the diameter of the circle is the radius of the semicircle.

4. The diameter of a semicircle is the length of one side of the right triangle.

5. The diagonal of the square is equal to the radius of the circle. Convert the square into a right triangle.

6. Rotate the inner square until its vertices touch the midpoints of the larger square. Observe that the radius of the outer circle is equal to the length of one side of the inner square.

7. Observe that the two shaded triangles are similar. Then use the fact that (ratio of areas) = (ratio of lengths)².

8. (a) Observe that triangles *AED* and *CED* share the same perpendicular height, or altitude.
   (b) Show that the area of triangle *BAD* is equal to the area of triangle *BCD*.

 9. Observe that the solid generated about each side of the right triangle is a cone.

10. If 1 cm on a map represents x km on the ground, then 1 cm² on the map will represent x^2 km².

Unit 5 (page 35)

1. Label the vertices of the shaded triangle, with base *AB* and the upper vertex *C*. Let *M* be the center of the hexagon. Then, triangle *AMB* would represent 1/6 of the area of the regular hexagon.

If the perpendicular height of triangle *AMB* has length r, then the perpendicular height (or altitude) of triangle *ACB* has length $2r$. Since both triangles have the same base, show that the area of triangle *ACB* is twice that of triangle *AMB*.

Unit 5 (page 35)

2. Relying on visual intuition to compare the sizes of different figures should be avoided —do the calculations.

3. Visually insert a square whose vertices touch the intersecting arcs. Shift some parts to see that the required area is the area of the square.

4. In a parallelogram, the sum of the interior angles is $180°$. Show that the sum of angles *CAB* and *CBA* is $60°$.

5. Shaded area = Area of sector – Area of triangle
   Use the Pythagorean theorem to find the length of the vertical chord, and show that the measure of the subtended angle is $120°$.

7. Observe that any tangent touching the circle is perpendicular to the radius.

8. Look for a pattern, by considering simple cases first.

9. (a) Ratio of areas = (Ratio of lengths)^2
   (b) Ratio of volumes = (Ratio of heights)^3
   (c) What is the volume of air in the cone?

10. Look out for any overlapping areas.

Unit 6 (page 44)

1. The most number of blocks that can be fitted along a given dimension of the box must be a multiple of the dimensions of each block—beware of any unused space.

2. The sum of the lengths any two sides in a triangle must be greater than the third side.

3. Try cutting one triangle up to make the other triangle. Or, use the Pythagorean theorem to find the perpendicular height of each triangle, before calculating the corresponding area.

4. Observe that the 12-gon is made up of 12 congruent triangles.

5. Rearrange the square into a right triangle.

7. Compare the cross-sectional areas of the pipes.

8. Draw the diagonals of the rectangle, and consider the sector and the triangle inside it.
   Shaded area = Area of sector – Area of triangle

9. Draw a vertical line of length, say, h, from X to the ground level, then use similar triangles.

10. How many diagonals are there in a 3-, 4-, 5-gon? Look for a pattern, then generalize the result for an n-gon.

Unit 7 (page 51)

1. Think of a diagonal.

2. Divide the triangle into two right triangles. Then use the Pythagorean theorem to find the perpendicular height (or altitude) common to both right triangles.

3. What is the exterior angle? The sum of the exterior angles in an *n*-gon is 360°.

4. How many types of each type of rectangle are there?

5. If the cube is of side *x* units, what is its surface area and volume?

6. How many square surfaces make up the total surface area of the solid? How many unit cubes make up the solid?

7. *Method 1*: The exterior angle in a triangle is the sum of its interior angles.
   Method 2: The angle subtended at the center is twice the angle subtended at the circumference.

8. Use the Pythagorean theorem to find *EF*, then use the trapezoid rule to find the area.

9. Observe that the diameter of circle *C* is perpendicular to the radius of circle *B*.

10. Imagine the figure being part of a bigger 4 by 4 square, then use a *look-see* proof to show that the figure is one-third of the bigger square.

Unit 8 (page 59)

1. Think of a right triangle.

2. Consider a simpler case.

3. Note that the two gears move in opposite directions. In a given time, the smaller gear will cover more revolutions, depending on the ratio of teeth between the two gears.

4. If *x* and *y* represent the sides of the right triangle formed as the ladder leans against the wall, and which also touches the edge of the box, use the Pythagorean theorem and similar triangles to formulate two equations connecting *x* and *y*.

5. Use the Pythagorean theorem to find the length of the hypotenuse, then use the formula for the area of a right triangle.

6. Show that the radius is $\sqrt{48}$ units. Then find half the length of the equilateral triangle.

7. A regular pentagon is a five-sided figure with equal length. Draw two diagonals, and show how that two of the triangles are similar.

Unit 8 (page 63)

8. Observe that the volume of the sphere that remains doesn't depend on the diameter of the hole.

9. Show that the sum of angles ABD and ACB is $180° - (p + q)$.

10. The sum of angles P, Q, and R is $180°$.

Unit 9 (page 67)

1. Observe that on simplifying, since x is an integer, for $13/y$ to be an integer, y must be 1 or 13.

2. Consider a simpler case, then look for a pattern.

3. Use the Pythagorean theorem to find the length of the diagonal, then complete the square to find its minimum value. *No calculus is needed here to find the smallest value.*

4. Draw a rectangle inside triangle XYZ, with one side being the perpendicular length from M to XZ.

5. Use the triangle's inequality twice.

6. (a) Express the length in terms of x.
   (b) Use the expression in (a) to complete the square. *No calculus is required.*

7. Express the square of the required length in terms of the surface areas.

8. Use the area of the triangle and the Pythagorean theorem.

9. Think of the right angle property in a semicircle.

10. Use dotted lines to find what fraction of triangle ABC represents square $BDEF$ and of square $PQRS$.

Unit 10 (page 75)

0. Imagine the inner circle becoming smaller and smaller until it shrinks to a point.

1. Consider a regular polygon inscribed in a circle, which is made up of isosceles triangles. Observe what happens when the bases of the triangles approach zero.

2. As the position of R changes, what happens to PR as R coincides with S?

3. What happens when the parallel lines come closer and closer to each other?

Unit 10 (page 77)

4. Since point P can lie anywhere between points A and B, what happens if P is moved to a limit, say, at A or B? What happens to triangle ABC then?

5. Consider a sphere made up of tightly packed cones.

6. Imagine the hill being so tiny that the campers spend a negligible time going up and down it. Or, assume that the campers spend the entire journey running up and down the hill, and almost no time walking on level ground. What is the the distance in each case?

7. See Worked Example 1.

8. What would happen if the smallest sphere for which a 6-cm tube were to be drilled through?

9. What would happen if the inner circle were to be made extremely small? What would happen to the radius then?

10. What would happen to its circumference if the radius of either planet were to be zero?

Unit 11 (page 83)

1. Sketch a 3D version of a $3 \times 3 \times 3$ cube, and count the most number of unit cubes that can be seen from any angle to ensure that no unit cubes are counted twice.

2. Make a systematic list for each type of square.

3. Show that the required difference is equivalent to the difference between the areas of the two circles.

4. The area of the larger square to be formed is equal to the sum of the areas of the two smaller squares.

5. In the small square, the shaded part represents 1 part, while the unshaded part, 4 parts. In the larger square, the shaded part represents 2 parts, while the unshaded part, 9 parts.

6. Try it out, beginning with a smaller number of points around the circumference.

7. In the smaller circle, 3 parts are shaded, and 2 parts are unshaded.

8. Imagine one vertical radius of the third circle touching the upper eyelid, and another radius touching the circumference of the second circle. Form a right triangle, one side being one of the radii and the other side being a horizontal radius of the second circle.

9. Any two tangents to a circle have equal length. Then use the Pythagorean theorem to find the diameter of the circle.

10. No solution exists if the smallest square measures 1 km by 1 km.

Unit 12 (page 90)

1. Observe that the tangent to the circle and the radius are perpendicular. Then use the Pythagorean theorem to find the radius.

2. Draw a perpendicular line from the origin O of the circle to meet the line, say, at D. Show that $OD = \sqrt{3}/2$. Observe that the radius OC is equal to the diagonal of the square, which is $\sqrt{2}$. In triangle OCD, use the Pythagorean theorem to show that $CD = \sqrt{5}/2$. Then find AC, before showing that $AC/AB = (1 + \sqrt{5})/2$.

3. (a) Show that $DE = \frac{1}{2} BC$.
   (b) Use the fact that when two chords of a circle intersect, the product of the segments of one chord is equal to the product of the segments of the other chord, i.e. $AE \times EC = ME \times EN$.

4. (a) Let x be equal to the given expression. Then $x = 1 + 1/x$. Solve for x.
   (b) Let y be equal to the given expression. Then $y = \sqrt{(1 + y)}$. Solve for y.

5. Show that $\sin 18° = 1/(\sqrt{5} + 1)$ or $1/2\Phi$; or $\cos 36° = \Phi/2$.

6. Substitute the linear equation into the quadratic equation.

7. Solve the two nonlinear equations simultaneously.

8. Use the fact that in a regular pentagon, every diagonal is parallel to the sides it does not intersect to establish the relationship: diagonal : side = side : (diagonal – side).

9. Use the fact that angles in the same segment are equal, and alternate angles are equal.

10.　(a) In a golden ratio triangle, side : base = Φ, the golden ratio.

Unit 13 (page 99)

For questions 1–10, use the heuristic, "Try it out" to get the required arrangement in each case.

Unit 14 (page 107)

1. Try it out with two circular cutouts.

2. Try it out using two identical coins.

3. Try it out with two circular cutouts.

4. Try it out using two circular cutouts.

5. Use the results of the SAT "Rolling Circle" question and question 4.

Unit 14 (page 110)

6. Try it out with two circular cutouts.

7. Imagine two circles, with one circumference being three times as long as the other one.

8. Try it out with coins or buttons.

9. Observe that the path is made up of three straight line segments and three circular arcs positioned at each of the three vertices. Observe that if the interior angle of the triangle is θ, then the "external" angle through the disc must rotate at the vertex is $\pi - \theta$.

10. The path consists of four straight line segments and four circular arcs at the vertices.

Unit 15 (page 114)

1. What is the formula for finding the interior angle of a regular pentagon?
   Interior angle of the star = 1/3 of the interior angle of the pentagon.

2. The angle subtended at the center is twice the angle subtended at the circumference.

3. Extend the line segments AC and EC to meet the circumference of the circle at points X and Y, respectively. Then use the fact that $\angle ACE = \frac{1}{2} \times (\angle XOY + \angle AOE)$.

4. Let the line segments EC and AC meet the circumference of the circle at points X and Y, respectively. Then use the fact that $\angle ACE = \frac{1}{2} \times (\angle AOE - \angle XOY)$.

5. Generalize the result for any star, so that its vertices each lie inside a circle. Then use the fact that the angle formed by two intersecting chords within a circle is half the sum of the arcs intercepted by the angle and its vertical angles.

6. Imagine all vertices of the star each lie inside a circle. Then use a method similar to solving question 5.

7. Use a similar method to solving question 6.

8. (a) Try it out.   (b) Try it out.

9. Try it out.

10. Shaded area of $\triangle ABC$: Unshaded area of $\triangle ABC$ = 2 : 1

Unit 16 (page 123)

1. Observe that $\angle BAM = \angle BMA$ and $\angle ABN = \angle ANB$.

2. If r is the radius of the circle, show that the area of the square is $(r\sqrt{2})^2$.

3. Area of triangle : Area of trapezoid = 1 : 4.

4. Label the square $ABCD$. Let the line segment through the center O of the circle parallel to AD meet AB at X. Then $OX = 1$. By the Pythagorean theorem, $1 + AX^2 = r^2$ and $XB = r$. Solve the equation $1 + (2 - r)^2 = r^2$.

5. List all the possible squares, especially those that are not horizontal and vertical.

6. Observe that the center of the inner circle and the points A and B form a right triangle.

7. Since all four parts share the same perpendicular height (or altitude), and since $QM = RM$, in terms of areas, $(A + B) : (C + D) = 1 : 1$.

8. Observe that triangles A and C are similar. Show that area $C = 4 \times$ area A. Also, note that triangles B and C have the same perpendicular height (or altitude) and that their bases are in the ratio 1 : 2.

9. Look for a pattern.

10. Apply the Pythagorean theorem to triangles ABE, BCE, and ACE.

Unit 17 (page 129)

1. Use a different method from the one provided in Unit 3, Question 7. Divide the right triangle into three smaller triangles, using the radius of the circle as their altitudes.

2. Use similar triangles to find the unknown length of the square.

6. Draw two equal right triangles of sides a and b. Then use the Pythagorean theorem to find their hypotenuses.

8. Let the square be of side 6 units. Show that the radius of each of the three lower circles is 1 unit. Observe that if the upper triangle is divided into two right triangles, each one is a 3-4-5 triangle. Show that the radius of each circle inside each right triangle is also 1 unit.

9. Imagine the two smaller semicircles being of the same size. What would be the shaded area?

Unit 18 (page 137)

2. (a) If *x* is the length of the square, use similar triangles to find the lengths of *BE* and *FC*. Then use the Pythagorean theorem to find *x*.

3. Label the sides of the cut-off right triangles by *a* and *b*, respectively. Show that $a^2 + b^2 = 128$. Then use the Pythagorean theorem to find the diagonal of the rectangle.

4. Show that $a + b = 90°$, then use the Pythagorean theorem to find *PR*.

5. Use the triangle's inequality to show that the only possible sides for the triangle are 2 cm, 3 cm, and 3 cm. Then use the Pythagorean theorem to find the height of the triangle.

6. If *r* and *R* are the radii of the two circles respectively, use the Pythagorean theorem to show that $R^2 - r^2 = 6^2$. Or, use the Zero Option to solve the question, by shifting the smaller semicircle until its center coincides with that of the larger semicircle, then imagine the two circles decreasing, with the chord remaining at 12 cm.

7. If *D* is the foot of the perpendicular height (or altitude) drawn from *C* to *AB*, then use the Pythagorean theorem to find the lengths of *AD* and *DB*.

8. (a) If the height of the 30-inch TV set is *x* inches, then its height is 1.8*x*. Use the Pythagorean theorem to find the value of *x*.
   (b) Compare the areas of the two TV sets.

9. If *r* is the radius of the pool, use the property of equal tangents to show that the hypothenuse is $(6 - r) + (8 - r)$. Then use the Pythagorean theorem to find the value of *r*.

10.   Think of two ways of finding the area of the triangle.

Unit 19 (page 144)

1. There are 4 ways of placing an arrow on any one side of the dice: N, S, E, W.

2. Try it out.

3. The sides of the square needn't always be horizontal and vertical.

4. Observe that halving the radius of the bigger circle leads to half of the bigger circle being enclosed. What happens when this process is repeated for the other smaller circles?

5. Find the perimeter and area of each figure.

7. Observe that each square pyramid fixed to each side of the cube contributes an extra of 4 edges. The number of edges in the new solid is the total number of edges of the cube plus the number of extra edges.

8. Visually see the cube being formed in your mind's eye.

Unit 19 (page 148)

9. Divide the figure into nine equal triangles, and see how many of these are unshaded.

10. Produce four breaks so that the five pieces could be used to form any number of cubes from 1 to 23.

Unit 20 (page 151)

1. Try it out.

2. Try it out, using everyday items, such as matchboxes.

3. (a) Investigate its area and perimeter.
   (b) If a and b are the adjacent sides of the right triangle, use the Pythagorean theorem, then show that to find whole numbers a and b, we need to solve the equation:
   $\frac{1}{2} ab = a + b + \sqrt{(a^2 + \underline{b}^2)}$.

4. Imagine the equilateral triangle being made up by three triangles formed by the line segments from Cain's position to each vertex of the triangle. Do the bases and their respective perpendicular heights (or altitudes) remain unchanged?

5. Sketch the possible triangles. How many different ones are there?

6. How many squares of each type are there?

7. If a and b are the length and the width of the rectangle, then we need to look for whole numbers a and b such that $ab = 2a + 2b$. Simplify to show that $a = 2b/(b - 2)$. What value(s) b can take so that a is a whole number?

8. Reflect the lower half of the figure along the vertical diameter to obtain three circles.

9. Try it out.

ANSWERS
&
SOLUTIONS

UNIT 1 (p. 3)

1. $\frac{1}{4}$.

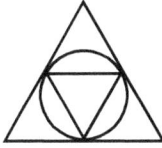

2. 2 units.

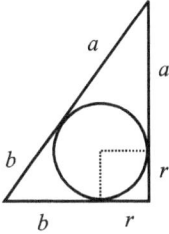

$2a + 2b + 2r = 6 + 8 + 10 = 24$

Since $a + b = 10$, $2a + 2b = 20$

Thus, $2r = 4$

Hence the radius, r, is 2 units.

3. $\frac{1}{5}$ square unit.

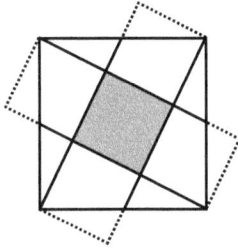

4. 4 units².

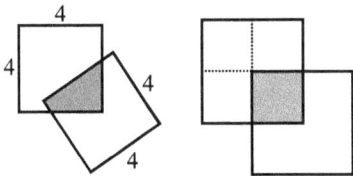

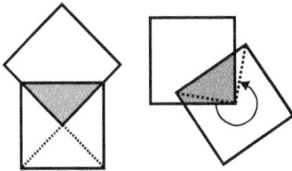

Shaded area = area of $\frac{1}{4}$ square

5. $x = 36$.

$ac = 45$, $ad = 15$, $bc = 108$

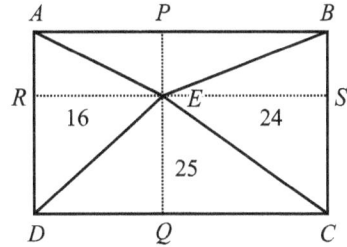

$x = bd = \dfrac{(bc)(ad)}{(ac)} = \dfrac{108 \times 15}{45} = 36$

6. $y = 15$.

Method 1

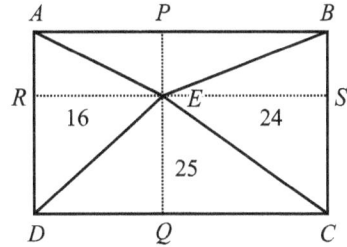

$AD = BC$

Area of $\triangle ADE = \dfrac{1}{2}(AD)(RE) = 16$

$\dfrac{1}{2}(AD)(DQ) = 16$ (since $RE = DQ$)

Area of rectangle $ADQP = (AD)(DQ) = 32$

Area of $\triangle BCE = \dfrac{1}{2}(BC)(ES)$

$= \dfrac{1}{2}(PQ)(QC)$

$= 24$ (since $BC = PQ$, $ES = QC$)

Area of rectangle $PQCB = (PQ)(QC) = 48$

Area of rectangle $ABCD = 32 + 48 = 80$

Therefore, shaded area, $y = 80 - (16 + 25 + 24)$

$= 15$

Method 2

$$\frac{h_1}{h_2}=\frac{16}{24}=\frac{2}{3}$$

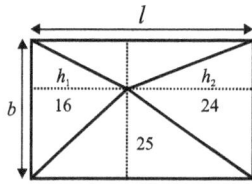

$$\Rightarrow h_1=\frac{2}{5}l$$

Also, $\frac{1}{2}bh_1=16$

$$\frac{1}{2}b(\frac{2}{5}l)=16$$

$$bl=80$$

$$y=80-(16+25+24)$$

$$=15$$

Method 3

In terms of areas, observe that

$$P+Q=U+V=a+b+c+d$$

$$U=P+Q-V$$

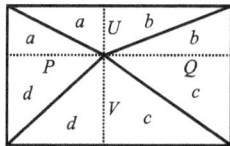

$$=16+24-25$$

$$=15$$

The value of y is 15.

7.   3 times.

In general, when a body rotates round a circle, it always makes one revolution more than one can count.

8.   Six slices.

The first cut slices the block in half.

Then put the two pieces on top of each other, and cut them into 4 slices.

The third cut gives 8 slices.

Then line up all 8 half-slices and cut them into 16 four-by-one slices.

The last two cuts will halve the pieces twice into 64 cubes.

9.   Not 90°, but 60°.

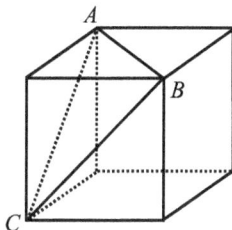

Since $AC = AB = BC$, the triangle ABC is equilateral.

Thus $\angle ABC = 60°$.

10.   72 units².

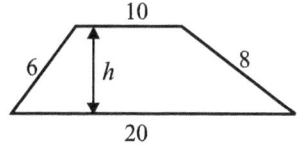

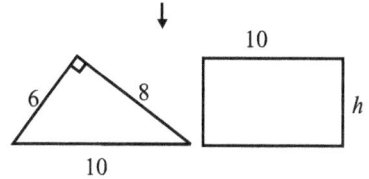

$$10 \times h = 6 \times 8$$

$$h = 4.8$$

Area of trapezium $= 1/2 \times 6 \times 8 + 10 \times h$
$$= 24 + 10 \times (4.8)$$
$$= 72$$

UNIT 2 (p. 11)

1.   1 : 2.

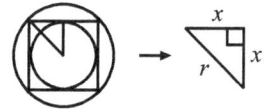

$$x^2 + x^2 = r^2$$

$$2x^2 = r^2$$

Ratio of the area of the smaller circle to the

area of the larger circle $= \dfrac{\pi x^2}{\pi r^2} = \dfrac{x^2}{r^2} = \dfrac{1}{2}$.

2.   20 cm.

Length of the smaller square

$$=\frac{68}{4} = 17 \text{ cm}$$

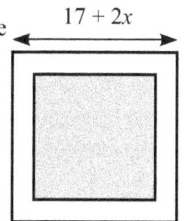

Length of the larger square
$$= 17 + 2x$$

$$(17 + 2x)^2 - 17^2 = 111$$
Solve for x.

3. 30 cm.

Let x be the length of one side of the equilateral triangle.

Then $(\frac{\sqrt{3}}{2}x)(\frac{x}{2}) = \frac{75}{\sqrt{3}}$

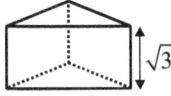

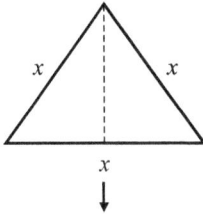

$\sqrt{x^2 - (\frac{x}{2})^2} = \frac{\sqrt{3}}{2}x$

Solve for x.

4. 17.

Ratio of areas = (ratio of lengths)^2

$= (\frac{\sqrt{51}}{\sqrt{3}})^2 = (\sqrt{\frac{51}{3}})^2 = \frac{51}{3} = 17.$

5. Let the radius of the small circle and the big circle be r and R, respectively.

Let each half of the chord XY be x.

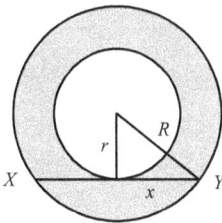

By the Pythagorean theorem, $x^2 + r^2 = R^2$

$x^2 = R^2 - r^2$

Shaded area $= \pi R^2 - \pi r^2$

$= \pi(R^2 - r^2)$

$= \pi(x^2)$

$= \pi(7^2) = 49\pi$

Alternatively, use the zero option technique to solve the problem. *See Unit 10.*

6.

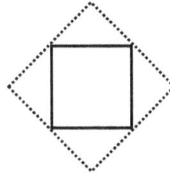

7. 125π cm^2.

If the two areas X and Y have a common area Z, then the unshaded areas are $(X - Z)$ and $(Y - Z)$.

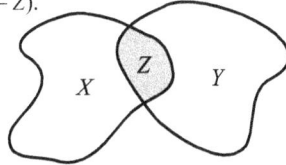

The difference of the areas of the unshaded regions is $[(X - Z) - (Y - Z)]$, i.e., $(X - Y)$. The difference is $\pi(15)^2 - \pi(10)^2 = 125\pi$.

8. 35 different triangles.

9. 28 cm^2.

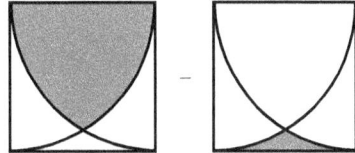

7 cm   7 cm

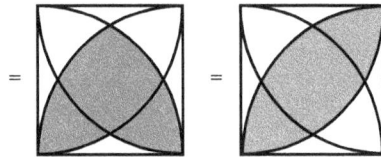

Difference between the two shaded areas

$= 2[(\frac{1}{4} \times \frac{22}{7} \times 7 \times 7) - \frac{1}{2} \times 7 \times 7]$

$= 28$ cm^2

10. $\frac{1}{6}$.

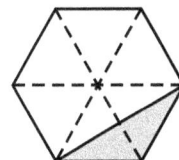

UNIT 3 (p. 20)

1. $\dfrac{\sqrt{2}-1}{\sqrt{2}+1}$ cm.

By Pythagoras' Theorem,

$OA = \sqrt{1^2 + 1^2} = \sqrt{2}$

$AC = OA - OC = \sqrt{2} - 1$

Radius, $r = AC - r.\sqrt{2}$

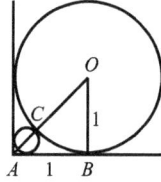

$r + r\sqrt{2} = \sqrt{2} - 1$

$r(1 + \sqrt{2}) = \sqrt{2} - 1$

$r = \dfrac{\sqrt{2}-1}{\sqrt{2}+1}$

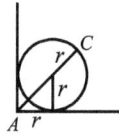

2. 3768 cm².

External surface area $= 2 \times (2\pi rh + \dfrac{1}{2} \times 2\pi rh)$.

3. 71.59 cm³.

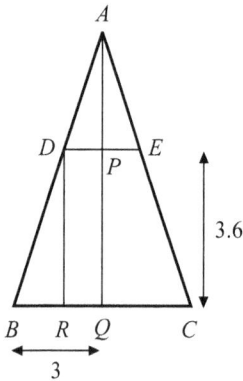

By similar triangles, $\dfrac{AP}{PD} = \dfrac{DR}{RB}$

Show that $AP = 7.2$ cm.

Height of the cone $= AQ = AP + 3.6$

Volume of the frustum

$=$ volume of larger cone

   $-$ volume of small cone cut off

4. Construct the squares, as shown in the diagram below.

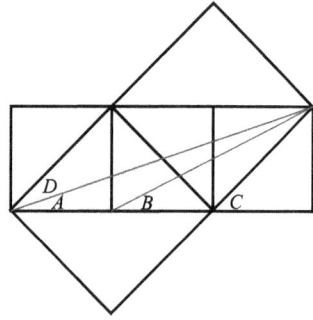

$\angle B = \angle D$ (corr. $\angle$s of similar right $\triangle$s)

$\angle A + \angle D = \angle C$

Thus $\angle C = \angle A + \angle B$

5. (a) $\dfrac{3}{4}$.

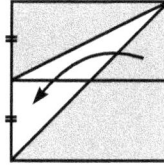

(b) $\dfrac{1}{6}$.

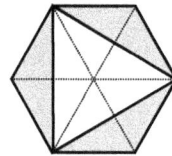

6. 25 feet.

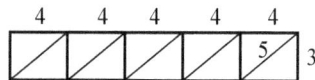

By the Pythagorean Theorem, $3^2 + 4^2 = 5^2$.
Since there are 5 complete spirals, the length of the spiral is $5 \times 5 = 25$ feet.

7. 1 m.

Let the radius be r m.

Clearly, $POQX$ is a square of side r m.

Now, $PZ = ZR = (3 - r)$ m

$QY = YR = (4 - r)$ m

Since $ZY = 5$ m

       $= ZR + YR$

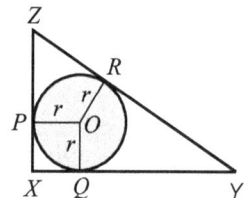

$(3 - r) + (4 - r) = 5$

$7 - 2r = 5$

$r = 1$

Therefore, the radius of the pool is 1 m.

8. $\dfrac{3}{2}$.

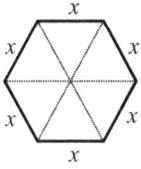 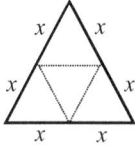

Hexagon Equilateral triangle

Perimeter of hexagon or triangle = $6x$ units

$$\frac{\text{Area of hexagon}}{\text{Area of triangle}} = \frac{6}{4} = \frac{3}{2}$$

9. *Hint:* Find whole numbers such that $l \times b = 96$ and $l + b = 20$.

10. 60 cm³.

If the cuboid has width of length l, b, and h cm,

then $lb = 20$

$lh = 15$

$bh = 12$

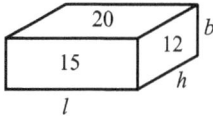

$(lb)(lh)(bh) = 20 \times 15 \times 12$

$(lbh)^2 = 20 \times 15 \times 12$

$lbh = \sqrt{20 \times 15 \times 12}$

$= 60$

Therefore, the volume of cuboid is 60 cm³.

Alternatively,

Find HCF (12, 15), HCF (12, 20), and HCF (15, 20).

Volume of cuboid

= HCF (12, 15) × HCF (12, 20) × HCF (15, 20)

= 3 × 4 × 5

= 60 cm³

UNIT 4 (p. 26)

1. 23 cubes.

2. 2 cm.

Let the smaller radius be x.

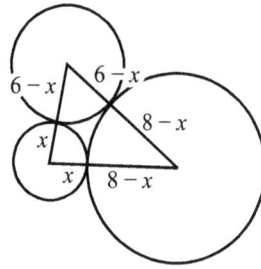

Then $(6 - x) + (8 - x) = 10$

$14 - 2x = 10$

$x = 2$

3. 1 : 1

Hint: Diameter of circle = radius of semicircle.

4. 48 units².

Observe that if

$\dfrac{1}{2}\pi r_1^2 = 9\pi$ and $\dfrac{1}{2}\pi r_2^2 = 16\pi$, then

Area $= \dfrac{1}{2} \times (2r_1) \times (2r_2)$

5. 10 cm².

Hint: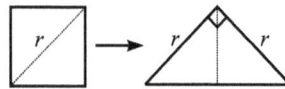

$\pi r^2 = 20\pi$

Area = 1/2 r^2

= 1/2 × 20

= 10

6. 5 cm.

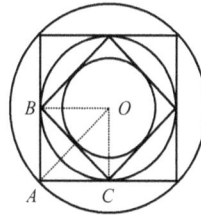

If we rotate the inner square 45°, as shown above, the radius of the large circle,

$OA = 5$ cm

= length of one side of the inner square, BC

(OA and BC are diagonals of the same square.)

Since the small circle is inscribed in the inner square, it has a diameter equal to the side of the square, BC, or 5 cm.

7. 98 cm².

Observe that the two shaded triangles are similar.

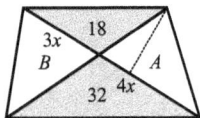

Ratio of their areas $= \dfrac{32}{18} = \dfrac{16}{9}$

Ratio of their lengths $= \sqrt{\dfrac{16}{9}} = \dfrac{4}{3}$

Observe that ΔA and the shaded smaller triangle have the same perpendicular height (or altitude), h.

Since their bases are in the ratio 4 : 3,

therefore, area of $\Delta A = \dfrac{4}{3} \times 18 = 24$ cm².

Using a similar argument, we can show that the area of ΔB is also 24 cm².

Hence, the area of the trapezoid is
$(18 + 32 + 24 + 24)$ cm² $= 98$ cm².

8. *Hint:*

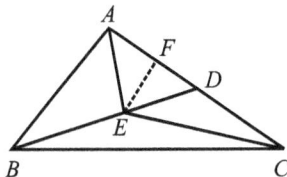

(a) Compare area of ΔAED and area of ΔCED with perpendicular height EF.

Note that $AD = DC$.

Observe that the median ED divides ΔAEC into two triangles having the same area.

(b) Use result in part (a).

Observe that in ΔBAC, BD is a median.

Show that area ΔBAD = area ΔBCD.

9. $\dfrac{3}{4}$.

Hint:

Observe that a cone is generated.

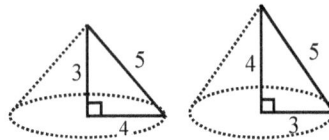

10. 1 : 250,000.

4 cm² represents 25 km²

2 cm represents 5 km

1 cm represents 5/2 km

Scale $= \dfrac{5/2 \times 1000 \times 100}{1}$ cm

UNIT 5 (p. 35)

1. $\dfrac{1}{3}$.

2. The inner circle is smaller.

Area of the outer circle $= \pi \times \left(\dfrac{6}{2}\right)^2 = 9\pi$

Area of the inner circle $= \pi \times \left(\dfrac{4}{2}\right)^2 = 4\pi$

Area of the space between the circles
$= 9\pi - 4\pi = 5\pi$

Therefore, the inner circle is smaller.

3. 8 cm².

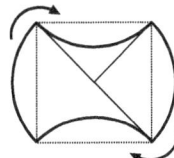

Required area = area of square

$= 2 \times$ area of triangle

$= 2 \times \dfrac{1}{2} \times 4 \times 2$ cm²

$= 8$ cm²

4.   120°.

$\angle A + \angle B = 180°$ (sum of int. $\angle$s)

$\angle CAB + \angle CBA = \dfrac{1}{3} \times 180° = 60°$

$\angle MCN = \angle ACB$ (vert. opp. $\angle$s)

$\quad\quad\quad\;\; = 180° - (\angle CAB + \angle CBA)$

$\quad\quad\quad\;\; = 180° - 60°$

$\quad\quad\quad\;\; = 120°$

5. $\dfrac{196}{3} - 49\sqrt{3}$ cm^2.

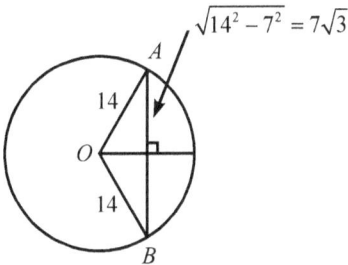

$\sqrt{14^2 - 7^2} = 7\sqrt{3}$

Show that $\angle AOB = 120°$.

6. $\dfrac{1}{4}$.

7.   6 units.

Hint:

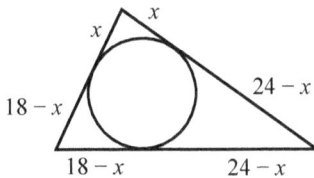

8.   27 triangles.

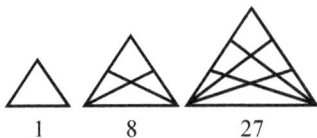

Observe the following pattern:

1	2	3	4	5
1^3	2^3	3^3	4^3	?

9.   (a) $\dfrac{1}{4}$.

$\dfrac{\text{Surface area } A}{\text{Surface area } B} = \left(\dfrac{\frac{1}{2}h}{h}\right)^2 = \dfrac{1}{4}$

(b) 60 cm^3.

$\dfrac{\text{Volume of water}}{\text{Total volume}} = (\text{Ratio of heights})^3$

$\dfrac{\text{Volume of water}}{480} = \left(\dfrac{1}{2}\right)^3$

$\text{Volume of water} = \dfrac{480}{8} = 60$ cm^3

(c) $L = \dfrac{\sqrt[3]{7}}{2}H$.

Volume of air in the cone

$= 480 - 60 = 420$ cm^3

$\left(\dfrac{L}{H}\right)^3 = \dfrac{420}{480}$

$L^3 = \dfrac{7}{8}H^3$

$L = \dfrac{\sqrt[3]{7}}{2}H$

10.   They are equal.

Hint:

The areas where they overlap take exactly the same area from the largest circle as they do from the smallest circles.

UNIT 6 (p. 44)

1.   Not 32, but 30 blocks.

The 5 edge fits into the 15 edge exactly 3 times.

The 7 edge fits into the 14 edge exactly 2 times.

The 3 edge fits inside the 16 edge 5 times with some left over.

The total number of blocks is $2 \times 3 \times 5 = 30$.

2. 5 units.

Since there are exactly two equal sides, the minimum perimeter cannot be $1 + 1 + 1 = 3$. Nor can it be $1 + 1 + 2 = 4$, because the longest side has to be less than the sum of the other two —here they are equal.
The smallest perimeter that fulfillsthegiven conditions is $1 + 2 + 2 = 5$.

3. They have the same area.

Hint: Cut one triangle to make the other triangle.

4. $3r^2$ square units.

The regular 12-sided polygon comprises 12 congruent triangles.

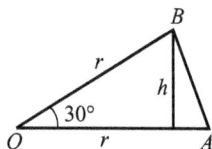

Area of $\triangle OAB =$

$$\frac{1}{2}bh = \frac{1}{2}r \times r\sin 30° = \frac{1}{2}r \times \frac{r}{2} = \frac{r^2}{4}$$
Area of polygon $= 12 \times \frac{r^2}{4} = 3r^2$

5. $\frac{1}{2}(x^2 + y^2)^2$.

$$\text{Area} = \frac{1}{2}d^2 = \frac{1}{2}(x^2 + y^2)^2$$

6. 4400 square units.
Rectangles $80 \times 55 = 110 \times 40$ will suffice.
The area of each rectangle is 4400.

7. 36 pipes.
Cross-sectional area of pipe $= \pi R^2$

Thus, $\dfrac{\text{large pipe}}{\text{small pipe}} = \dfrac{6^2}{1^2} = 36$

The large pipe has the carrying capacity of 36 small pipes.

8. $\left(\dfrac{\pi}{3} - \dfrac{\sqrt{3}}{4}\right)$ cm².

Hint:
Angle of the sector
$= 120°$
Use the Pythagorean Theorem to find the length of rectangle.

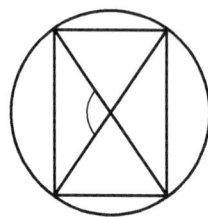

Shaded area = Area of sector – Area of triangle

9. $2\dfrac{2}{5}$ m.

$$\frac{4}{a+b} = \frac{h}{b}$$
$$\frac{6}{a+b} = \frac{h}{a}$$
$$h = \frac{4b}{a+b} = \frac{6a}{a+b}$$
$$4b = 6a$$
$$b = \frac{3}{2}a$$
$$h = \frac{6a}{\frac{5}{2}a} = 2\frac{2}{5}$$

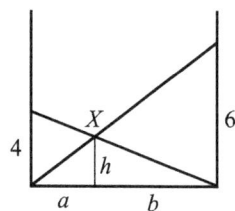

10. 4850 diagonals.

$$d = \frac{n(n-3)}{2} = \frac{100 \times 97}{2} = 4850$$

UNIT 7 (p. 51)

1. *Hint:* Think of a diamond.

The diagonal of the square is the side of the bigger square.

2. 84 square units.

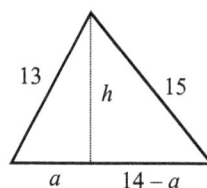

178

$$a^2 + h^2 = 13^2 \quad (1)$$
$$h^2 + (14-a)^2 = 15^2 \quad (2)$$

From (2), $h^2 + 14^2 - 28a + a^2 = 15^2$
$$13^2 + 14^2 - 28a = 15^2$$
$$28a = 140$$
$$a = 5$$

$$h^2 = 13^2 - 5^2$$
$$= (13+5)(13-5)$$
$$= 18 \times 8$$
$$= 9 \times 16$$
$$= 3^2 \times 4^2$$

$$A = \frac{1}{2} \times 14 \times h$$
$$= \frac{1}{2} \times 14 \times 3 \times 4$$
$$= 84$$

3. 9 sides.

A regular polygon can be inscribed in a circle.

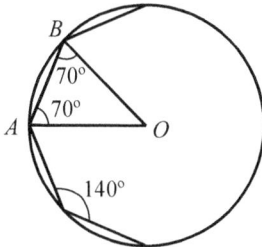

If OA and OB are the radii, then $\angle AOB = 40°$.
Thus, there are $\dfrac{360}{40} = 9$ such angles around point O. Hence, the polygon has nine sides.

Alternatively,
Interior angle = 140°
Exterior angle = 180° − 140° = 40°
Sum of exterior angles of an n-gon = 360°
Thus, $40n = 360$
$$n = 9$$
Hence the polygon has 9 sides.

4. 90 rectangles.

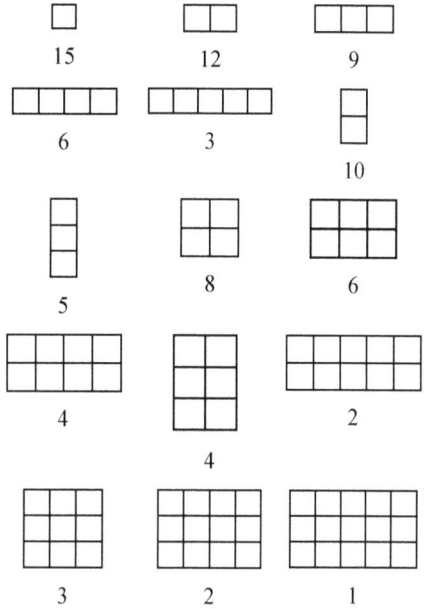

5. 216 cubic units.

$$x^3 = 6x =$$
$$x^2(x-6) = 0$$
$$x^2 = 0 \text{ or } x = 6$$
Since $x > 0$, $x = 6$
Volume of cube = $6 \times 6 \times 6 = 216$

6. 144 cm³.

There are altogether 48 (= 18 + 12 + 9 + 9) squares, which make up the total surface area.

48 squares yield 192 cm²
Area of one square = 4 cm²
Length of one side = 2 cm
Volume of solid = volume of 18 cubes
$$= 18 \times 8 \text{ cm}^3$$
$$= 144 \text{ cm}^3$$

7. 180°.

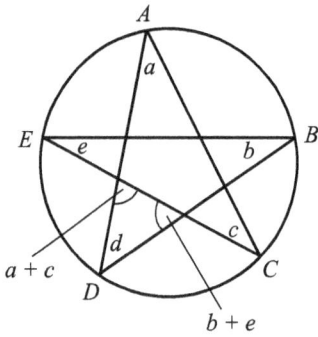

Alternatively,

Each inscribed angle measures one-half the measure of the arc that subtends it.

Since the five inscribed angles are subtended by five arcs that measure a total of 360°,

the measures of the five inscribed angles is

$(\frac{1}{2} \times 360°) = 180°$.

8. $\frac{13}{32}$ square unit.

By Pythagorean theorem, $0.5^2 + (1-x)^2 = x^2$

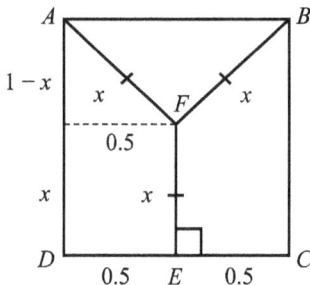

Show that $x = \frac{5}{8}$, and use the trapezoid rule

to find the required area.

9. The area of the circular track equals the area of circle C.

By Pythagoras' theorem, $r_C^2 + r_B^2 = r_A^2$

Area of circular track $= \pi r_A^2 - \pi r_B^2$

$= \pi r_C^2$

$= $ area of circle C

10. Short-circuiting the use of the formula for a geometric series:

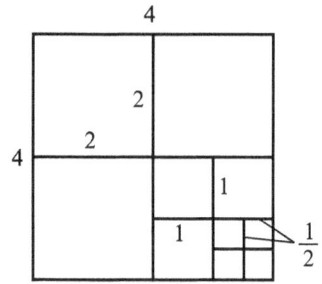

$3(2^2 + 1^2 + (\frac{1}{2})^2 + ...) = 16$

Area of figure $= \dfrac{16}{3}$ square units

UNIT 8 (p. 59)

1. 50 cm.

The angle between the two given sides must be 90 degrees.

Observe that a triangle with sides in the ratio 3 : 4 : 5 is a right triangle.

2. 17 snaps.

3. (a) −16 revolutions
(b) +6 revolutions.

Hint:

If gear *A* revolves *x* times per minute, gear *B* revolves −2*x* times per minute, where the negative indicates the opposite direction.

4. 3.62 m or 1.38 m.

Let the legs of the right triangle formed be *x* and *y*.

By the Pythagorean theorem, $x^2 + y^2 = 15$ (1)

By similar triangles, $\dfrac{x-1}{1} = \dfrac{1}{y-1}$, or $xy = x + y$

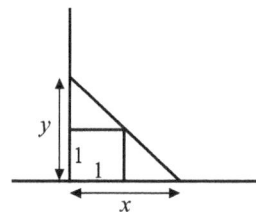

Then $x^2 + y^2 + 2xy = (x + y)^2$

$$= 15 + 2(x + y)$$

$$(x + y)^2 - 2(x + y) - 15 = 0$$

$$[(x + y) - 5] [(x + y) + 3] = 0$$

$$(x + y) = 5 \text{ or } (x + y) = -3$$

Since $x + y > 0$, we have

$$x + y = 5$$

or

$$y = 5 - x \quad (2)$$

Substituting (2) into (1), we have

$$x^2 + (5 - x)^2 = 15$$

Simplifying, $x^2 - 5x + 5 = 0$

Solving, $x \approx 3.62$ or 1.38

Therefore, the ladder touches the wall at 3.62 m.

5. $(3 - 2\sqrt{2})a^2$.

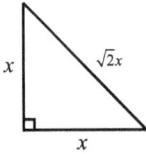

If x is one of the equal sides, then

$$2x + \sqrt{2}x = 2a$$

$$x(2 + \sqrt{2}) = 2a$$

$$x = \frac{2a}{2 + \sqrt{2}}$$

$$\text{Area} = \frac{1}{2}x^2$$

$$= \frac{1}{2}\left(\frac{2a}{2 + \sqrt{2}}\right)^2$$

$$= \frac{2a^2}{6 + 4\sqrt{2}}$$

$$= \frac{a^2(3 - 2\sqrt{2})}{(3 + 2\sqrt{2})(3 - 2\sqrt{2})}$$

$$= (3 - 2\sqrt{2})a^2$$

6. 72 units.

Show that one side of the triangle is 24 units.

7. $\dfrac{1 + \sqrt{5}}{2}$ (≈ 1.618).

Hint: Show that $\triangle ABC$ and $\triangle BDC$ are similar.

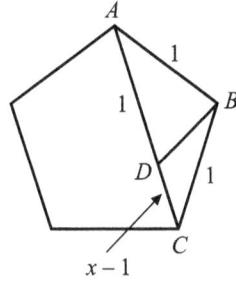

If the diagonal AC is x, show that $x = \dfrac{1 + \sqrt{5}}{2}$.

8. 36π cm^3.

Hint:

The volume of the sphere that remains is always the same as the volume of a solid sphere whose diameter is equal to the hole's length.

The volume of the sphere that remains after the hole is drilled is independent of the hole's diameter.

If we reduce the hole's diameter to zero, the hole becomes a straight line that is the diameter of a solid sphere.

$$\text{Volume} = \frac{4}{3}\pi(3^3) = 36\pi \text{ cm}^3$$

9. $q = 3p - 180°$.

Hint:

$\triangle ABD = 180° - 2x = \triangle ACB$

$\triangle ADB = \triangle CAD + \triangle ACB$.

10. $1 : \sqrt{3} : 2$.

Hint:

Use Sine Rule.

Also $\angle P + \angle Q + \angle R = 180°$

UNIT 9 (p. 67)

1. 169 cm².

$$xy + y = y^2 + 13$$

$$x = \frac{13}{y} + y - 1$$

Observe that, since x is an integer, $y = 1$ or 13.

If $y = 1$, $x = \frac{13}{1} + 1 - 1 = 13$

If $y = 13$, $x = \frac{13}{13} + 13 - 1 = 13$

Maximum area of rectangle = $13 \times 13 = 169$ cm²

2. 30 squares.

$16 + 9 + 4 + 1 = 30$

3. 8 cm.

Let x and $8\sqrt{2} - x$ be the length and width of the rectangle, respectively.

If d is the diagonal of the rectangle, then by the Pythagorean Theorem,

$$d^2 = x^2 + (8\sqrt{2} - x)^2$$
$$d^2 = 2x^2 - 16\sqrt{2}x + 128$$

Using the method of completing the square, we have

$$d^2 = 2[(x - 4\sqrt{2})^2 + 32]$$

d^2 is minimum when $x = 4\sqrt{2}$.

Therefore, the minimum value of d is $\sqrt{2(0^2 + 32)} = 8$.

4. 24 square units.

Hint:

Observe that $MN = 3$.

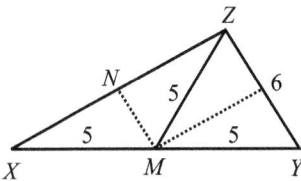

By Pythagoras' Theorem, $XN = 4$.

Then $XZ = 8$.

5. 5993 cm.

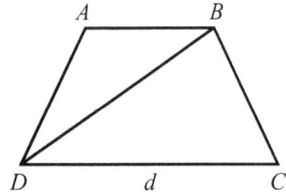

Let $ABCD$ represent the quadrilateral and $DC = d$ cm.

By the triangle inequality,

$d = DC < BC + BD < BC + AB + AD$

$\qquad\qquad = 1997 + 1998 + 1999$

$\qquad\qquad = 5994$

Thus, $d \leq 5993$.

6. (a) $A = x(80 - 2x)$.
   (b) 20 m.

Use completing the square to show that
$$A = -2(x - 20)^2 + 800.$$

7. 9 cm

If the cuboid measures a cm by b cm by c cm, with $a \leq b \leq c$, then

$$bc = 72, \quad ac = 54, \quad ab = 48$$

$$c^2 = \frac{ac \times bc}{ab}$$
$$= \frac{72 \times 54}{48}$$
$$= \frac{(9 \times 8) \times (6 \times 9)}{48}$$
$$= 9 \times 9$$
$$c = 9$$

Hence the length of the longest edge is 9 cm.

8. 75 cm.

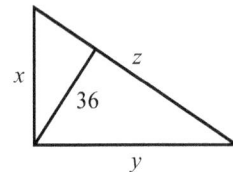

$x + y + z = 180$

$x + y = 180 - z \qquad\qquad (1)$

Area of triangle $= \frac{1}{2} z \times 36 = \frac{1}{2} xy \qquad (2)$

By the Pythagorean Theorem, $x^2 + y^2 = z^2$ (3)

Now $x^2 + y^2 + 2xy = (x + y)^2$

$\qquad\qquad\qquad = z^2 + 72z$ [from (2) & (3)]

From (1), $x + y = 180 - z$

$\qquad \sqrt{z^2 + 72z} = 180 - z$

Simplifying, $z = 75$

9.   8.

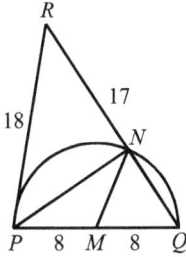

In $\triangle PQN$ with $\angle N = 90°$, the midpoint M of PQ is the circumcentre of $\triangle PQN$.

Thus, $PM = QM = MN = 8$ (right angle property in semicircle)

10. $\dfrac{8}{9}$.

Method 1

Let $AB = 2$ units

$\triangle FAE$ and $\triangle FBE$ are isosceles right-angled triangles.

$\therefore AF = FE = FB$

Now, $AF + FB = AB$

$= 2$ units.

$\therefore BF = 1$ unit

$\therefore$ Area of square

$BDEF = 1$ square unit

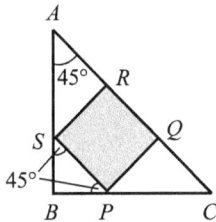

Let $BS = x$

$PS = \dfrac{x}{\sin 45°} = \sqrt{2}x$

$\qquad = RS$ ($PQRS$ is a square)

$AS = \dfrac{RS}{\sin 45°} = \sqrt{2}RS$

$\qquad\quad = 2x$ ($\triangle RSA$ is an isosceles

$\qquad\qquad\qquad$ right-angled $\triangle$)

$\therefore 2 = AB = AS + BS$

$\qquad\quad = 2x + x$

$\qquad\quad = 3x$

$\qquad x = \dfrac{2}{3}$

Area of square $PSRQ = RS^2$

$\qquad\qquad\qquad = (\sqrt{2}x)^2$

$\qquad\qquad\qquad = (\dfrac{2}{3}\sqrt{2})^2$

$\qquad\qquad\qquad = \dfrac{8}{9}$

So, $\dfrac{\text{area of } PQRS}{\text{area of } BDEF} = \dfrac{\frac{8}{9}}{1} = \dfrac{8}{9}$

Method 2

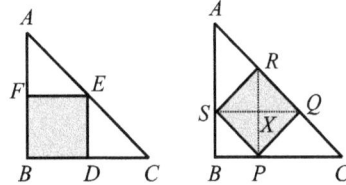

Let the area of $\triangle AFE$ be a units².

Then $\triangle AFE$, $\triangle BFE$, $\triangle BDE$ and $\triangle CDE$ are all congruent.

So, $\dfrac{\text{area of square } BDEF}{\text{area of } \triangle ABC} = \dfrac{2a}{4a} = \dfrac{1}{2}$

$\triangle ASQ$ is an isosceles right triangle, so $\triangle ARS \equiv \triangle QRS$

Similarly $\triangle CQP \equiv \triangle RQP$.

Let RP and QS meet at X, and let $\triangle XSP$ have area b units².

Then area of square $PQRS = 4b$

$\triangle BPS \equiv \triangle XSP$, and area of $\triangle ARS = 2b = $ area of $\triangle CQP$

Hence, area $PQRS$: area of $\triangle ABC = \dfrac{4b}{9b} = \dfrac{4}{9}$

Method 3

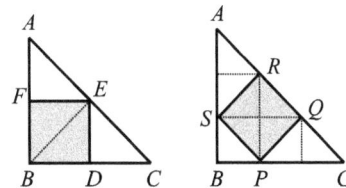

Area of $BDEF = \dfrac{2}{4}$ area of $\triangle ABC$

Area of $PQRS = \dfrac{4}{9}$ area of $\triangle ABC$

$\dfrac{\text{Area of square } PQRS}{\text{Area of square } BDEF} = \dfrac{\frac{4}{9}}{\frac{2}{4}} = \dfrac{4}{9} \times \dfrac{4}{2} = \dfrac{8}{9}$

UNIT 10 (p. 75)

1. 20π cm.

   Let the smaller circle become smaller and smaller until it shrinks to a point.

   Then the circle reduced to a point would become the center of the larger circle.

   The distance between the two circles would then become the radius of the larger circle. The difference between the lengths of the circumferences of the two circles is now the circumference of the larger circle, or 20π cm.

2. Consider a polygon inscribed in a circle, being divided into isosceles triangles.

   Let b_1, b_2, b_3, ... be the bases of the triangles, and h the height of each triangle.

   Then the area of the polygon is

   $\frac{h}{2}(b_1 + b_2 + b_3 + \cdots + b_n)$.

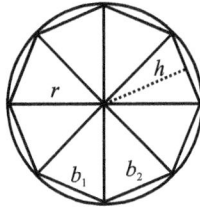

   Applying the zero option, we make the bases of the triangles approach zero.

   The sum of all the bases becomes the circumference of the circle, and h becomes the radius, r.

   As a result, the area of the circle is

   $$\frac{r}{2} \times 2\pi r = \pi r^2.$$

2. 10 units.

   Observe that PR remains the same length regardless of the position of R.

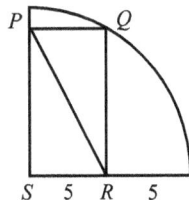

   When R coincides with S, PR becomes the radius.

   Hence the diagonal PR is 10 units long.

Alternatively,

The line PR is one diagonal of the rectangle $PQRS$.

The other diagonal, QS, is the radius of the circle, which is 10 units.

Since $PR = QS$, line PR is also 10 units long.

3. When parallel lines are cut by a transversal, alternate interior angles are congruent.

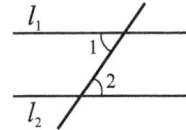

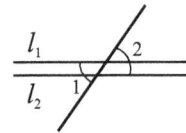

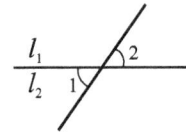

   Since l_1 and l_2 are parallel, they appear to merge into a single line, thus making the alternate interior angles appear to be vertical lines.

4. 10 cm.

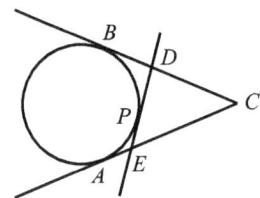

   Since P can lie anywhere on the circle from A to B, we move P to a limit, either at A or B.

   In either case, one side of $\triangle ABC$ reduces to zero as side DE expands to 5 cm, thus producing a straight-line triangle with sides of 5, 5 and 0.

   Hence the perimeter is 10 cm.

Alternatively,

Since lines tangent to a circle from an exterior point are equal, $AE = EP$ and $PD = DB$.

Since $DE = DP + PE$, the perimeter of the triangle must be $5 + 5 = 10$ cm.

184

5. Consider a sphere completely filled with tightly packed cones.

Then the volume of the sphere can be approximated by volume of cone 1 + volume of cone 2 +···+ volume of cone n.

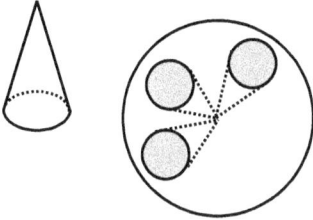

If we let the bases of the cones become progressively smaller, the heights of the cones approach the radius of the sphere, and the sum of the bases of the cones approaches the surface of the sphere.

Thus the volume of the sphere is

$$\frac{1}{3} \times 4\pi r^2 \times r = \frac{4}{3}\pi r^3.$$

6. 24 km.

Time taken walking = 12 00 − 6.00 = 6 h

Let the hill be so tiny that the campers spend a negligible time going up and down it.

Then they practically walk the entire journey at 4 km/h.

So in 6 h they travel a total distance of 6 × 4 = 24 km.

Now, assume the campers spend practically the entire journey running up and down the hill, and almost no time walking on the level.

If the hill is x km long, then their walking time is $\frac{x}{3} + \frac{x}{6} = 6$

$$x = 1$$

Therefore, the total distance is 2 × 12 = 24 km again.

The campers walk at the same average speed on the level as when they walk up and down the hill. The time they lose walking uphill they recover walking downhill, since $\frac{1}{3} + \frac{1}{6} = 2 \times \frac{1}{4}$

Hence any combination of hill and level road yields the same result.

Alternatively,

Let the distance on level ground and uphill/downhill be x km and y km, respectively.

Then $2(\frac{x}{4}) + (\frac{y}{3} + \frac{y}{6}) = 6$

$$3x + 3y = 36$$
$$x + y = 12$$

Distance walked = $2x + 2y$
$$= 24 \text{ km}$$

7. The area of the ring is 36π.

8. $36\pi \text{ cm}^3$.

Determine the smallest sphere for which a 6-cm tube could be drilled through.

This approaches a sphere whose diameter is 6 cm, for which the width of the tube approaches 0.

If the volume of the resulting ring is the same for all tubes of 6-cm length, then the ring volume is equal to the volume of a sphere 6 cm in diameter, that is, $V = \frac{\pi \times 6^3}{6} = 36\pi$.

9. We are looking for the circumference of these circles.

The size of neither circle is given. Suppose the small (inner) circle is extremely small, so small that it has a radius of length 0 and it is thus reduced to a point. Then the distance between the circles becomes the radius of the larger circle.

The circumference of this larger circle is $2\pi R = C + 1$, where C is the circumference of the earth (now reduced to 0) and $C + 1$ is the length of the rope.

$$2\pi R = 0 + 1 = 1$$

The distance between the circles is $R = 1/2\pi = 0.159$ meters, which would allow a mouse to comfortably fit beneath the rope.

10. Assume the radius of the earth or moon, and therefore its circumference to be zero.

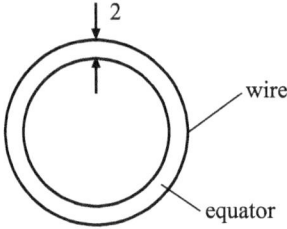

We are then left with a circle of radius 2 meters, or 4π meters.

Algebraically,

Consider a sphere of radius R meters with the wire along its surface.

Length of the wire $= 2\pi R$

With the wire raised at a distance 2 meters above the surface, the new radius is $(R + 2)$ meters.

Length of the circular wire $= 2\pi \times (R + 2)$

$= (2\pi R + 4\pi)$ m

So the difference in length is 4π meters, which is independent of R.

Therefore, regardless whether it is the earth or moon, the extra length of wire will be the same, i.e., 4π meters.

UNIT 11 (p. 83)

1. 19 cubes.

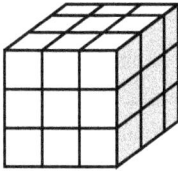

2. 6 different squares.

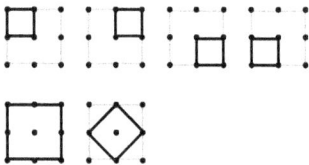

3. 7π cm^2.

4. 20 cm.

5. 2 : 17.

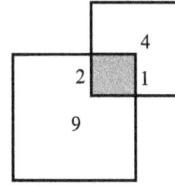

Smaller square	:	Shaded area	:	Larger square

$\left.\begin{matrix}4\\8\end{matrix}\right) \times 2$: $\left.\begin{matrix}1\\2\end{matrix}\right) \times 2$

2 : 9

Ratio of shaded area to unshaded area

$= 2 : 8 + 9$

$= 2 : 17$

6. 17 chords.

7. $\dfrac{3}{20}$.

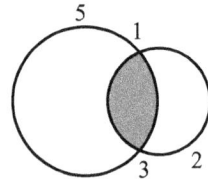

Smaller circle	:	Shaded area	:	Larger circle
5	:	1		
15	:	3		
		3	:	2

Fraction of the figure that is shaded $= \dfrac{3}{15 + 3 + 2}$

$= \dfrac{3}{20}$

8. 5 units.

If the radius of the small circle is 1 unit, then the radius of the second circle is 3 units.

If r is the radius of the big circle, then

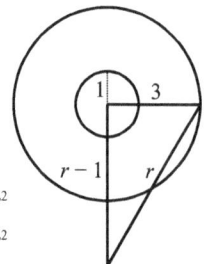

$3^2 + (r - 1)^2 = r^2$

$9 + r^2 - 2r + 1 = r^2$

$2r = 10$

$r = 5$

186

9. Any two tangents to a circle have equal length, so we can segment the length as follows:

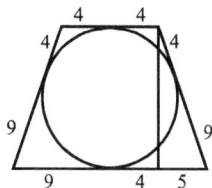

The diameter of the circle is the leg of the right triangle with hypotenuse 13 cm. By the Pythagorean theorem, the diameter is given $\sqrt{13^2 - 5^2} = 12$

Hence the radius of the circle is 6 cm.

10. There is no solution if the smallest square is only 1 km on each side. However, a solution exists when the smallest square has a side of 2 km.

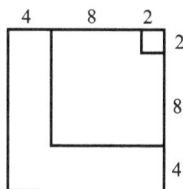

The plots of the second and third sons each measure 96 km², and the total area of the farmer's plot was 196 km².

UNIT 12 (p. 90)

1. $\dfrac{\sqrt{5}-1}{2}$ cm, or 0.618 cm (to 3 sig. fig.)

2. $\dfrac{1+\sqrt{5}}{2}$.

3. (a) Draw a line segment through E parallel to AB and intersecting BC at F. The resulting $BDEF$ is a rhombus with each side of length 1 unit.

   (b) Let $MD = x = EN$

   Using the fact that when two chords of a circle intersect, the product of the segments of one chord equals the product of the segments of the other chord, we have

   $$AE \times EC = ME \times EN$$
   $$1 \times 1 = (x+1)x$$
   $$x^2 + x - 1 = 0$$
   $$x = \frac{-1 \pm \sqrt{1-4(1)(1)}}{2(1)} = \frac{-1 \pm \sqrt{5}}{2}$$

Since $x > 0$, $x = \dfrac{\sqrt{5}-1}{2}$

Hence, $\dfrac{MD}{DE} = \dfrac{\sqrt{5}-1}{2}$ or $\dfrac{EN}{DE} = \dfrac{\sqrt{5}-1}{2}$

6. At $\left(\dfrac{1+\sqrt{5}}{2}, \dfrac{1+\sqrt{5}}{2}\right)$ and $\left(\dfrac{1-\sqrt{5}}{2}, \dfrac{1-\sqrt{5}}{2}\right)$

   or (ϕ, ϕ) and $\left(-\dfrac{1}{\phi}, -\dfrac{1}{\phi}\right)$

7. At $\left(\phi, \dfrac{1}{\phi}\right)$, $\left(-\phi, -\dfrac{1}{\phi}\right)$, $\left(\dfrac{1}{\phi}, \phi\right)$, $\left(-\dfrac{1}{\phi}, -\phi\right)$,

   where $\phi = \dfrac{1+\sqrt{5}}{2}$ and $\dfrac{1}{\phi} = \dfrac{\sqrt{5}-1}{2}$

8. To show that the ratio of diagonal of a regular pentagon to its side length is the golden ratio, ϕ, we use the fact that in a regular pentagon every diagonal is parallel to the sides it does not intersect.

   Since $\triangle AED$ and $\triangle BTC$ have parallel sides, they are similar to each other.

   $\therefore \dfrac{AD}{AE} = \dfrac{BC}{BT}$

   But $BT = BD - TD$
   $\quad\quad\quad = BD - AE$

   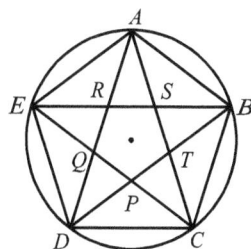

   In the regular pentagon,

   diagonal : side = side : (diagonal − side)

   $$\frac{d}{a} = \frac{a}{d-a} \quad \text{or} \quad \frac{d}{a} = \frac{1}{\dfrac{d}{a}-1}$$

   where d is the length of the diagonal and a is the length of the side.

   Let $x = \dfrac{d}{a}$, then we get the equation

   $$x = \frac{1}{x-1}$$
   $$x^2 - x - 1 = 0$$

   Since $x > 0$, $x = \dfrac{\sqrt{5}+1}{2}$

   $$\frac{d}{a} = \frac{\sqrt{5}+1}{2}$$

9. 140°.

$20° = \angle ABE$ (given)

$\quad = \angle ACE$ ($\angle$s in same segment)

$\quad = \angle DEC$ ($AC \parallel ED$)

$\quad = \angle DAC$ ($\angle$s in same segment)

$\therefore \angle AXC = 180° - (\angle XAC + \angle XCA) = 140°$

10. For each golden ratio triangle, $\dfrac{\text{side}}{\text{base}} = \phi$

$$\therefore \dfrac{AD}{DC} = \phi$$

Since $DC = 1$, $AD = \phi$.

In golden $\triangle AEH$,

$\dfrac{\text{side}}{\text{base}} = \phi = \dfrac{AE}{AH}$

$\therefore AH = \dfrac{1}{\phi}$

Since $EH = DC = 1$,

then $FH = EH - EF$

$\quad = 1 - \dfrac{1}{\phi}$

$\quad = \dfrac{1}{\phi^2}$ (since $\phi^2 - \phi - 1 = 0$,

$\quad$ or $\quad 1 - \dfrac{1}{\phi} = \dfrac{1}{\phi^2}$)

UNIT 13 (p. 99)

1.

2.

3.

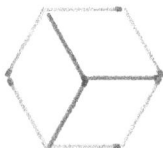

4.

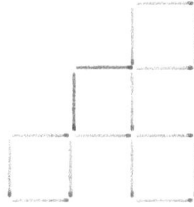

5.

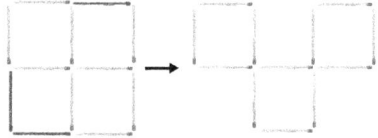

6.

7.

8.

9.

10.

UNIT 14 (p. 107)

1. *A* makes 3 revolutions.

2. *B* makes 2 revolutions.

3. *A* makes one and half revolutions.

4. The number of revolutions is two. Here are their positions:

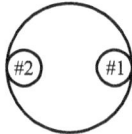

The "inside" problem applies to planetary gear trains.

5. Two times.
   The outer sphere rotates 4 times, whereas the inner sphere 2 times.

6. The point *P* travels along a fixed diameter of *B*. All of the circular motion is somehow transformed into linear motion.

7. Four times.
   In general, when a body rotates around a circle, it always makes one revolution more than one can count.

8. Six circles.
   Observe that the inscribed triangle *ABC* is equilateral, since each side equals the length of two radii.

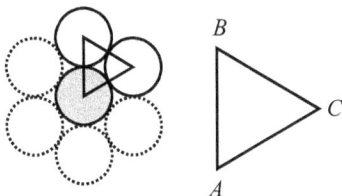

Since each angle in an equilateral triangle is 60°, angle *A* is 60°, which is one-sixth of a circle.

9. $4(\pi + 3)$ cm.
   Hint: The path traced by the center of the disc is made up of two separate parts:
   (i) 3 straight line segments,
   (ii) 3 circular arc centered at each of the three vertices.

10. $2(\pi + 10)$ cm.

UNIT 15 (p. 114)

1. 180°.

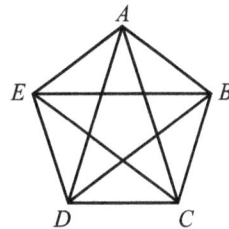

Since *ABCDE* is a regular pentagon,

$$\angle EAB = 180° - \frac{360°}{5} \quad \text{or} \quad \frac{3 \times 180°}{5}$$
$$= 180° - 72° \quad \text{or} = 3 \times 36°$$
$$= 108° \qquad\qquad = 108°$$

$$\angle DAC = \frac{108°}{3} = 36°$$

Sum of the interior angles of the star
$= 5 \times 36° = 180°$

2. 180°.

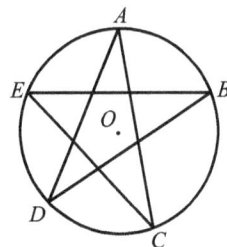

If *O* is the center of the circle, then

$$\angle CAD + \angle DBE + \angle ECA + \angle ADB + \angle BEC$$
$$= \frac{\angle COD}{2} + \frac{\angle DOE}{2} + \frac{\angle EOA}{2} + \frac{\angle AOB}{2} + \frac{\angle BOC}{2}$$
($\angle$ at center = 2 $\angle$ at circumference)
$$= \frac{1}{2}(\angle COD + \angle DOE + \angle EOA + \angle AOB + \angle BOC)$$
$$= \frac{1}{2} \times 360°$$
$$= 180°$$

3. 180°.

Let O be the center of the circle.

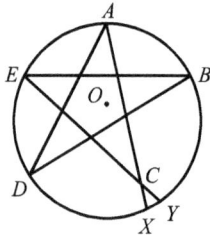

$$\angle ACE = \frac{1}{2}(\angle XOY + \angle AOE)$$

($\angle$ formed by 2 intersecting chords within a

circle $= \frac{1}{2} \times$ sum of the arcs intercepted by the

angle and its vertical angles)

Sum of the internal angles of star

$$= \angle CAD + \angle DBE + \angle ECA + \angle ADB + \angle BEC$$
$$= \frac{1}{2}XOD + \frac{1}{2}\angle DOE + \frac{1}{2}(\angle XOY + \angle AOE)$$
$$\quad + \frac{1}{2}\angle AOB + \frac{1}{2}\angle BOC$$
$$= \frac{1}{2}(\angle XOD + \angle DOE + \angle XOY + \angle AOE$$
$$\quad + \angle AOB + \angle BOC)$$
$$= \frac{1}{2} \times 360°$$
$$= 180°$$

4. 180°.

Let O be the center of the circle.

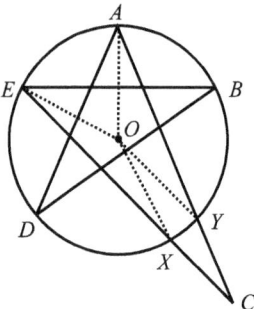

$$\angle ACE = \frac{1}{2}(\angle AOE - \angle XOY)$$

Sum of the interior angles of the star

$$= \angle CAD + \angle DBE + \angle ECA + \angle ADB + \angle BEC$$
$$= \frac{1}{2}\angle DOY + \frac{1}{2}\angle DOE + \frac{1}{2}(\angle EOA - \angle XOY)$$
$$\quad + \frac{1}{2}\angle AOB + \frac{1}{2}\angle BOX$$
$$= \frac{1}{2}(\angle DOY + \angle DOE + \angle EOA - \angle XOY$$
$$\quad + \angle AOB + \angle BOX)$$
$$= \frac{1}{2} \times 360°$$
$$= 180°$$

5. For any star, a circle may be built so that the star's vertices lie inside it.

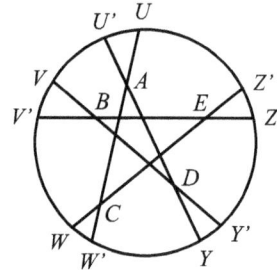

If O is the center of the circle, then

$$\angle DAC = \frac{1}{2}(\angle UOU' + \angle W'OY)$$
$$\angle EBD = \frac{1}{2}(\angle VOV' + \angle Y'OZ)$$
$$\angle ACE = \frac{1}{2}(\angle WOW' + \angle Z'OU)$$
$$\angle BDA = \frac{1}{2}(\angle YOY' + \angle U'OV)$$
$$\angle CEB = \frac{1}{2}(\angle ZOZ' + \angle V'OW)$$

Therefore,

$$\angle DAC + \angle EBD + \angle ACE + \angle BDA + \angle CEB$$
$$= \frac{360°}{2} = 180°$$

6. 180°.

Hint: Imagine the star's vertices lying inside a circle. Then use a similar method to question 5.

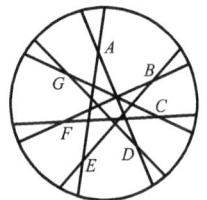

7. 540°.

Hint: Use a similar method to question 6.

8. (a) No matter how hard we try, a six-pointed star cannot be completed in this way.
   (b) Connect point 1 with 3 with 5 with 2 with 4 with 1,

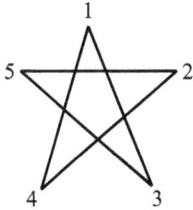

9. There are two seven-pointed stars:

   (i) skipping one point at a time, connect point 1 with 3 with 5 with 7 with 2 with 4 with 6 with 1; or

   Note that the stars is based on the fact that 7 is coprime with 2 (and 5).

   (ii) skipping two points at a time, connect point 1 with 4 with 7 with 3 with 6 with 2 with 5 with 1.

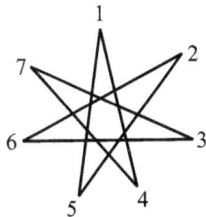

10. 1 : 1.

$$\frac{\text{Shaded area of}}{\triangle ABC} : \frac{\text{Unshaded area of}}{\triangle ABC} = 2 : 1$$

$$\frac{\text{Shaded area of}}{\triangle PQR} : \frac{\text{Unshaded area of}}{\triangle PQR} = 2 : 1$$

Since shaded area of $\triangle ABC$ = shaded area of $\triangle PQR$,

shaded area of $\triangle ABC$ or $\triangle PQR$

: unshaded area of $\triangle ABC$ + $\triangle PQR$

= 2 : 2

= 1 : 1

UNIT 16 (p. 123)

1. 110°.

   Hint: Show that, in $\triangle ABM$, $\angle B = 35°$ and in $\triangle ABP$, $\angle A = 35°$.

   Then $\angle P = 180° - 70° = 110°$.

2. $\frac{2}{\pi}$.

   Let's take a two-dimensional cross-section of the round hole and the square peg.

   If the hole has radius r, then the square peg has cross-section of side $r\sqrt{2}$.

   Fraction of hole occupied by peg

   $$= \frac{\text{area of square}}{\text{area of circle}}$$

   $$= \frac{(r\sqrt{2})^2}{\pi r^2}$$

   $$= \frac{2}{\pi}$$

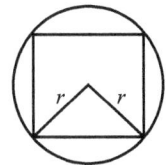

3. 2 : 3.

   Let the area of figure BEA be 1 square unit.

   Then the area of figure $CDAE$ will be 4 square units.

   Area of figure $CDFE$

   = area of $CDAE$ − area FAE

   = 3 square units

   $a : b$ = area of $EFAB$: area of $CDFE$

   = 2 : 3

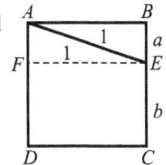

   Alternatively,

   Let the length of the rectangle be w.

   Then the area of the triangle is $\frac{1}{2} \times a \times w$.

   The area of the trapezoid $ADCE$ with parallel sides AD and EC and lengths $a + b$ and b respectively, has area $\frac{(2b+a) \times w}{2}$

   Now, $\frac{aw}{2} : \frac{(2b+a)w}{2} = 1 : 4$

   Simplify the above to show that

   $\frac{a}{b} = \frac{2}{3}$.

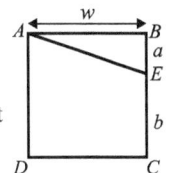

4. $\frac{5}{4}$ cm.

5. 11 squares.

5 small   4 medium   2 large

6. π cm².

The center of the circle and points A and B form a right triangle.
Let the radius of the inner circle be r_1 and the radius of the outer circle be r_2.

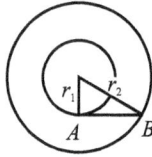

By the Pythagorean theorem,
$r_2^2 - r_1^2 = 1$

Area of annular region $= \pi(r_2^2 - r_1^2)$
$$= \pi \times 1$$
$$= \pi \text{ cm}^2$$

7. 24 : 25.

$A : B = 3 : 2; C : D = 7 : 5$
Since $QM = RM$, in terms of areas,
$(A + B) : (C + D) = 1 : 1$

Since $A + B = 3 + 2 = 5$ units, $C + D = 5$ units

$B = 2$ units, $D = \frac{5}{12} \times 5$
$$= \frac{25}{12} \text{ units}$$

$\dfrac{B}{D} = \dfrac{2}{\frac{25}{12}}$

$\qquad = 2 \times \dfrac{12}{25}$

$\qquad = \dfrac{24}{25}$

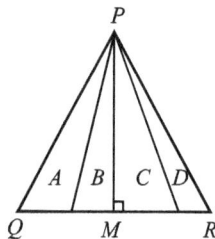

The ratio of area B to area D is 24 : 25.

8. $A = 5$ cm², $B = 10$ cm², $C = 20$ cm², $D = 25$ cm².

Notice that triangles A and C are similar triangles.
Since their sides are 1 : 2, their areas are 1 : 4.
Thus, area $C = 4 \times$ area A

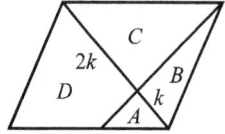

Consider triangles B and C.
Both have the same height and their bases are in the ratio 1 : 2, so their areas are 1 : 2. Thus, area $C = 2 \times$ area B

This means that $C = 4A$ and $B = 2A$

Due to symmetry, $D + A = B + C = 2A + 4A$

So, $D = 5A$
Therefore, $A : B : C : D = 1 : 2 : 4 : 5$

Since the area of the parallelogram is 60 cm², the respective areas are:

$A = 5$ cm², $B = 10$ cm², $C = 20$ cm², $D = 25$ cm²

Alternatively,
Area of $\Delta B = 10$ cm²
Area of $\Delta C = 20$ cm²

$\therefore$ Area of $\Delta A = \dfrac{1}{2}$ area of $\Delta B = 5$ cm²
Area of figure $D = 60 - 5 - 10 - 20$
$$= 25 \text{ cm}^2$$

9. 204 squares.

Type of squares	Number
1 × 1	64
2 × 2	49
3 × 3	36
4 × 4	25
5 × 5	16
6 × 6	9
7 × 7	4
8 × 8	1
Total	204

In general, on a $n \times n$ grid, the total number of squares of all sizes is the sum of the squares of all the numbers from 1 to n.

10.   41 cm and 50 cm.

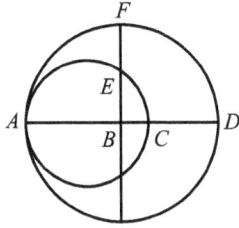

Let the radius of the larger circle be r cm.

Then $BC = r - 9$ cm and $BE = r - 5$ cm

Thus, $r - 5$ is the geometric mean of $r - 9$ and r, from which $r = 25$ cm.

The diameters of the circles are thus 50 cm and 41 cm respectively.

Alternatively, apply the Pythagorean Theorem to right triangles ABE, BCE, and ACE.

UNIT 17 (p. 129)

1.   The radius is 1 unit.

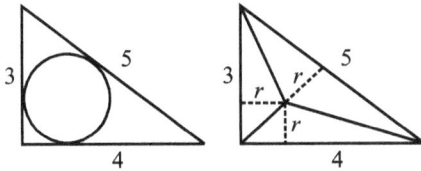

Area of triangle $= \dfrac{1}{2} \times 3 \times 4 = 6$

Each small triangle has height r, so their areas are

$$\frac{1}{2} \times 3r, \; \frac{1}{2} \times 4r \text{ and } \frac{1}{2} \times 5r.$$

Since the sum of the areas of the smaller triangles is 6 units, we have

$$6 = \frac{1}{2} \times 3r + 2r + \frac{1}{2} \times 5r = 6r$$
$$r = 1$$

2. $1\dfrac{5}{7}$ m.

By similar triangles,

$$\frac{3-x}{x} = \frac{3}{4}$$

Solve for x.

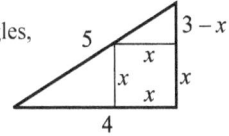

3. $\dfrac{4}{9}$ cm.

6.   Since the bottom sides of the squares form a segment of length $a + b$ that can be divided into segments of length b and a, we draw two equal right triangles of sides a and b.

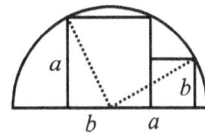

By Pythagoras' theorem, the hypotenuses of the two triangles are $\sqrt{a^2 + b^2}$, and are perpendicular to each other. Therefore the common vertex of these triangles is the center of the circle and its radius is $\sqrt{a^2 + b^2}$.

7.   The other radii are 3 cm, 2 cm and 1 cm. Observe that the centers of the circles and semicircles form various 3-4-5 triangles, and that the figure can be constructed by drawing the small circle first, before circumscribing the large one.

8.   If the square has side 6 units, we have a 3-4-5 triangle, and we need to prove that the radius of its inscribed circle is 1 unit.

9.   The shaded area is equal to the area of a circle that has a diameter of l.

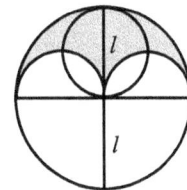

Archimedes also proved that if two circles are inscribed on either side of line l, touching it, they are equal.

193

UNIT 18 (p. 137)

1. 1430 units.

$550 = 5 \times 110$

$1320 = 12 \times 110$

By the Pythagorean Theorem,

$$\text{length of hypotenuse} = \sqrt{550^2 + 1320^2}$$
$$= \sqrt{110 \times (5^2 + 12^2)}$$
$$= 110 \times 13$$
$$= 1430 \text{ units}$$

2. 15/2 cm².

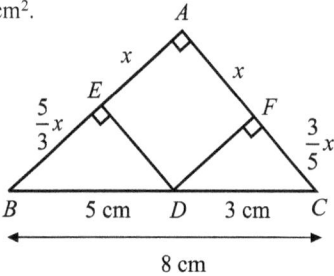

If x is the side of square $AFDE$, use similar triangles to show that:

Area of $\triangle ABC = \dfrac{32}{15}x^2$

Area of triangles BED and $DFC = \dfrac{17}{15}x^2$

3. 16 cm.

The total area of the cut-off pieces is

$$2 \times \frac{1}{2}a^2 + 2 \times \frac{1}{2}b^2$$
$$= a^2 + b^2$$
$$= 128$$

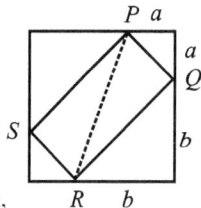

By Pythagorean Theorem,

$$PR^2 = PQ^2 + QR^2$$
$$= (a^2 + a^2) + (b^2 + b^2)$$
$$= 2(a^2 + b^2)$$
$$= 2 \times 128$$
$$= 256$$
$$PR = \sqrt{256} = 16$$

The length of the diagonal of the rectangle is 16 cm.

4. 24 cm².

If $\angle PRM = a$ and $\angle QRM = b$,
then since $PM = RM = MQ$,
we have $\angle RPQ = a$ and $\angle RQP = b$

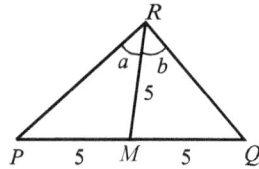

In $\triangle PQR$,

$a + (a + b) + b = 180°$ ($\angle$sum in a $\triangle = 180°$)

$a + b = 90°$

Thus, $\angle PRQ = 90°$

By Pythagorean Theorem,

$$PR = \sqrt{PQ^2 - QR^2} = \sqrt{10^2 - 6^2} = 8$$

Area of $\triangle PQR = \dfrac{1}{2} \times QR \times PR$

$$= \frac{1}{2} \times 6 \times 8$$
$$= 24 \text{ cm}^2$$

5. If a, b and c are the lengths of the three sides of the triangle, then
   (a) $a + b + c = 8$, and
   (b) $a + b > c$, $a + c > b$ and
       $b + c > a$ (by triangle inequality)

Checking values for a, b and c, the only possible solution is $a = 2$, $b = 3$, $c = 3$. This implies that the triangle is isosceles.

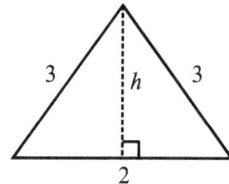

By Pythagorean Theorem,

$$h = \sqrt{3^2 - 1^2}$$
$$= \sqrt{8}$$

Area of triangle $= \dfrac{1}{2} \times 2 \times \sqrt{8}$

$$= \sqrt{8}$$
$$= 2\sqrt{2} \text{ cm}^2$$

6. 18π cm^2.

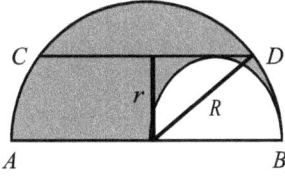

Let r and R be the radii of the two circles.
The perpendicular line from the center of the semicircle to the chord CD will bisect the chord.
Since $CD = 12$ cm, by Pythagorean Theorem,
$$R^2 - r^2 = 6^2$$

Area of shaded region $= \dfrac{1}{2}(\pi R^2 - \pi r^2)$

$\qquad\qquad\qquad\quad = \dfrac{1}{2}\pi(R^2 - r^2)$

$\qquad\qquad\qquad\quad = \dfrac{1}{2}\pi \times 6^2$

$\qquad\qquad\qquad\quad = 18\pi$ cm^2

Alternatively,

The above inelegant solution uses the Pythagoras' theorem.

Instead, let's solve the problem intuitively.

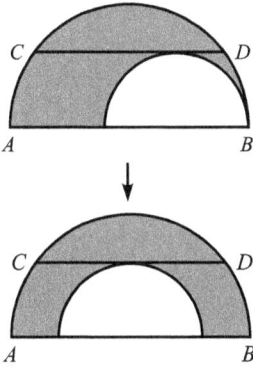

Shift the smaller semicircle so that its center coincides with the center of the larger circle. Imagine the two circles decreasing, with the chord remaining at 12 cm, until the smaller circle becomes a point circle. Its radius would then become zero, and what is left is a circle whose center is the larger circle and whose diameter is the chord CD.

Area of new circle $= \dfrac{1}{2}\pi r^2$

$\qquad\qquad\qquad\quad = \dfrac{1}{2}\pi(6)^2$

$\qquad\qquad\qquad\quad = 18\pi$ cm^2

7. 84 units^2.

Let D be the foot of the perpendicular height drawn from C to AB.

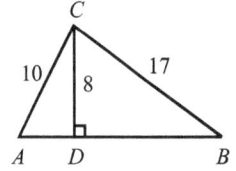

Then $AD = \sqrt{10^2 - 8^2} = 6$
and
$$DB = \sqrt{17^2 - 8^2} = 15$$

So, $AB = AD + DB$

$\qquad\quad = 6 + 15$

$\qquad\quad = 21$

Area of $\triangle ABC = \dfrac{1}{2} \times 8 \times 21$

$\qquad\qquad\qquad\quad = 84$ units^2

8. (a) About 14.57 inches.
    (b) 212.3 square inches.

(a) Let the height of the 30-inch set be x.
     Then its width is $1.8x$.
     By Pythagoras' Theorem,
     $$x^2 + (1.8x)^2 = 30^2$$
     $$4.24x^2 = 900$$
     $$x \approx 14.57 \text{ inches}$$

(b) If the 20-inch set has height x and width $1.8x$, then
     $$x^2 + (1.8x)^2 = 20^2$$
     $$4.24x^2 = 400$$
     $$x \approx 9.71 \text{ inches}$$
     $$1.8x = 17.48 \text{ inches}$$
     The dimensions of the 20-inch TV are approximately 9.71 inches by 17.48 inches.
     Area of the 30-inch TV
     $\approx 14.57 \times 26.22$
     ≈ 382.03 square inches
     Area of the 20-inch TV
     $\approx 9.71 \times 17.48$
     ≈ 169.73 square inches

Difference in areas $\approx 382.03 - 169.73$
$\qquad\qquad\qquad\qquad = 169.73$ square inches

Note:
The area of the 30-inch TV is over twice the area of the 20-inch TV even though the ratio of the diagonals is only 1.5 : 1.

9.   2 m.

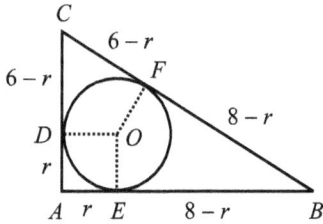

Let r be the radius of the pool.

$AEOD$ is a square, of side r m.

$CD = CF = 6 - r$

$BE = BF = 8 - r$

By Pythagoras' theorem, $BC = \sqrt{6^2 + 8^2} = 10$ m

$$6 - r + 8 - r = 10$$
$$14 - 2r = 10$$
$$r = 2$$

Therefore the radius of the pool is 2 m.

10. $\dfrac{qr}{\sqrt{q^2 + r^2}}$.

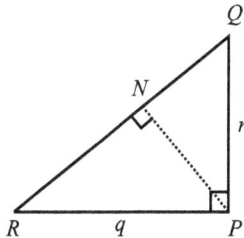

Area of $PQR = \dfrac{1}{2}qr$

But it is also equal to $\dfrac{1}{2} \times QR \times PN$

$$= \dfrac{1}{2}\sqrt{q^2 + r^2} \times PN$$

Thus, $\dfrac{1}{2}qr = \dfrac{1}{2}\sqrt{q^2 + r^2} \times PN$

$$PN = \dfrac{qr}{\sqrt{q^2 + r^2}}$$

UNIT 19 (p. 144)

1.   There are 4 ways of placing an arrow on any one side of the dice.

So one can place the arrows in $4 \times 4 \times 4 \times 4 \times 4 \times 4 = 4096$ different ways.

If we consider the symmetries of a cube, which means eliminating configurations that are symmetric duplicates leaves us with only 192 different ways to label the cube with arrows.

2.

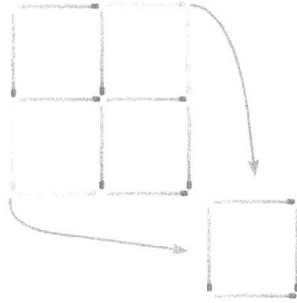

3.   (a) 5 sizes.

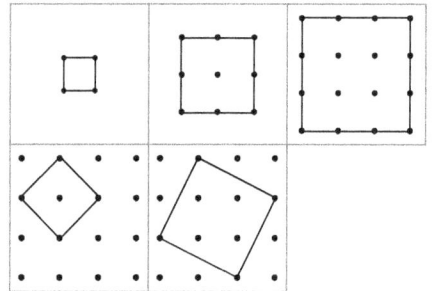

(b) 20 squares ($9 + 4 + 1 + 4 + 2$)

4.   The second larger circle is half the radius of the bigger circle, so therefore it has one quarter of the area. However, there are two such circles, so together they enclose half the area of the bigger circle—including all the enclosed circles they enclose. The same argument can be repeated for all the other circles.

5. Their perimeters equal their areas.

Area of triangle $= \dfrac{1}{2} \times 6 \times 8 = 24$ square units

Perimeter of triangle $= 6 + 8 + 10 = 24$ units

Area of circle $= \pi \times r^2 = 4\pi$ square units

Perimeter of circle $= 2 \times \pi \times 2 = 4\pi$ units

Area of square $= 4 \times 4 = 16$ square units

Perimeter of square $= 4 + 4 + 4 + 4 = 16$ units

6. Three reasons why round is the best possible shape:
   1. Round manhole covers cannot fall through their round holes accidentally.

      Square and other polygonal covers can.
   2. Heavy round covers can be rolled into position, while other shapes would have to be carried.
   3. Round covers can cover holes no matter how they are oriented relative to the hole. Square holes fit only when they are positioned in one of four orientations.

7. 36 edges.

   There are 12 edges in a cube.

   To each of the 6 faces are added 4 edges, which means a total of 24 edges.

   The total number of edges is $12 + 24 = 36$.

8. 120 $(= 4 \times 5 \times 6)$.

9. $\dfrac{5}{9}$.

   Five out of nine equal parts are unshaded.

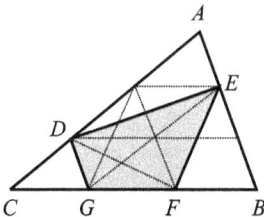

10. Four breaks that produce 1 cube, 2 cubes, 3 cubes, 7 cubes and 10 cubes enables us to obtain any number of cubes from 1 to 23. So do 4 breaks that produce 1 cube, 1 cube, 3 cubes, 6 cubes and 12 cubes.

1, 2, 3, 7, 10	
1 = 1	14 = 1 + 3 + 10
2 = 2	15 = 2 + 3 + 10
3 = 3	16 = 1 + 2 + 3 + 10
4 = 1 + 3	17 = 7 + 10
5 = 2 + 3	18 = 1 + 7 + 10
6 = 1 + 2 + 3	19 = 2 + 7 + 10
7 = 7	20 = 3 + 7 + 10
8 = 1 + 7	21 = 1 + 3 + 7 + 10
9 = 2 + 7	22 = 2 + 3 + 7 + 10
10 = 10	23 = 1 + 2 + 3 + 7 + 10
11 = 1 + 10	
12 = 2 + 10	
13 = 3 + 10	

1, 1, 3, 6, 12	
1 = 1	13 = 1 + 12
2 = 1 + 1	14 = 1 + 1 + 12
3 = 3	15 = 3 + 12
4 = 1 + 3	16 = 1 + 3 + 12
5 = 1 + 1 + 3	17 = 1 + 1 + 3 + 12
6 = 6	18 = 6 + 12
7 = 1 + 6	19 = 1 + 6 + 12
8 = 1 + 1 + 6	20 = 1 + 1 + 6 + 12
9 = 3 + 6	21 = 3 + 6 + 12
10 = 1 + 3 + 6	22 = 1 + 3 + 6 + 12
11 = 1 + 1 + 3 + 6	23 = 1 + 1 + 3 + 6 + 12
12 = 12	

UNIT 20 (p. 151)

1.

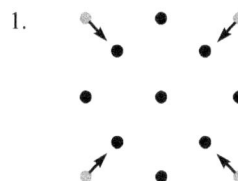

2. Place the bricks, as shown in the figure below.

   With the ruler, measure the distance between points A and B, which is the desired length.

   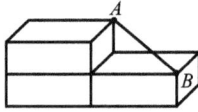

3. (a) Its perimeter equals its area.

   (b) 5-12-13

   We need to find whole numbers a and b such that

   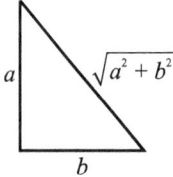

   $$\frac{1}{2}ab = a + b + \sqrt{a^2 + b^2}$$

   $$\frac{1}{2}ab - a - b = \sqrt{a^2 + b^2}$$

   $$[ab - (2a + 2b)]^2 = 4(a^2 + b^2)$$

   Simplifying, we have
   $$ab - 4a - 4b + 8 = 0$$
   $$b(a - 4) = 4a - 8$$
   $$b = \frac{4a - 8}{a - 4}$$

   $a = 5, b = 12$, so
   $$\sqrt{a^2 + b^2} = \sqrt{5^2 + 12^2} = 13$$

   $P = 12 + 5 + 12 = 30$,
   $$A = \frac{1}{2} \times 5 \times 12 = 30$$

   For $a = 6, b = 8$, we have the given triangle.

4. Cain is right.

   Area of triangle = Sum of the areas of the three triangles formed by the lines from Cain to the corners

   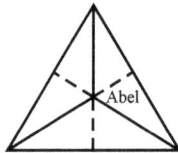

   This area (which is fixed) depends on the lengths of the sides (which are fixed) and the sum of the perpendicular distances (which therefore must also be fixed).

5. 3 triangles.

   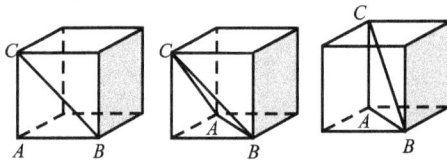

6. 46 squares.

7. Area = 18 square units
   Perimeter = $2 \times (6 + 3) = 18$ units

   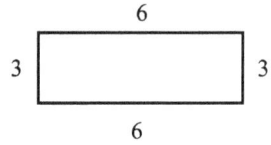

8. 2 cm².

   Reflect the lower half of the figure in the vertical diameter to yield three 'kissing' circles.

   Since the ratio of their radii is $1 : 2 : 3$, the ratio of their areas is $1 : 4 : 9$.

   So the shaded area is

   $$\frac{4 - 1}{9} \times \text{area of circle}$$

   $$= \frac{3}{9} \times 6 = 2 \text{ cm}^2$$

9. They always make a rectangle because the bases add up to 180°, so the angles made by their bisectors add up to 90°. Each point where their bisectors cross is the third angle of the triangle, the other two angles of which add up to 90°, so this angle is also 90°.

   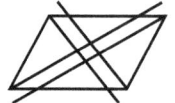

10. By the ratio of areas of similar figures, the smaller semicircle is one quarter of the area of the larger. So the small semicircle is equal in area to either of the two sectors, in particular the upper one. So, since

    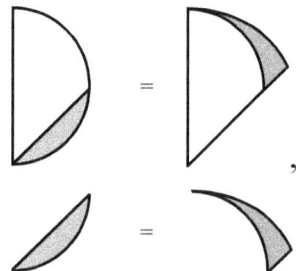

Geometrical
Quickies & Trickies
ONLINE

For more *quickies*, *trickies*, and *toughies*, visit the facebook page: **fb.com/singaporemathplus**

Bibliography & References

Acheson, D. (2002). *1089 and all that: A journey into mathematics*. Oxford: Oxford University Press.

Andrews, Colin & Spignesi, Stephen J. (2003). *Crop circles: Signs of contact.* Frankilin Lakes, NJ: Career Press.

Benson, S. *et al.* (2005). *Ways to think about mathematics:* Activities and investigations for grade 6-12 teachers. Thousand Oaks, California: Corwin Press.

Birtwistle, C. (1971). *Mathematical puzzles and perplexities: How to make the most of them.* London: George Allen & Unwin Ltd.

Bodycombe, D. J. (1996). *The mammoth book of brainstorming puzzles.* London: Robinson Publishing.

Book, D. L. (1992). *Problems for puzzlebusters.* Washington, DC: Enigmatics Press.

Bradis, V. M., Minkovskii, V. L. & Kharcheva, A. K. (1999). *Lapses in mathematical reasoning.* Moneola, New York: Dover Publications, Inc.

Brecher, E. (1994). *The ultimate book of puzzles, mathematical diversions, and brainteasers.* London: Pan Books.

Darling, D. (2004). *The universal book of mathematics.* New Jersey: Castle Books.

De Villiers, M. (1999). Stars: A second look. *Mathematics in School,* 28(5), 30.

Eves, Howard (2001). *Mathematical reminiscences.* The Mathematical Association of America.

Fauconnier, Gilles & Turner, Mark (2002). *The way we think: Conceptual blending and the mind's hidden complexities.* Basic Books.

Fetisov, A. I. & Dubnov, Y. S. (2006). *Proof in geometry with "miskakes in geometric proofs."* Moneola, New York: Dover Publications, Inc.

Gardiner, A. (1987). *Mathematical puzzling.* New York: Dover Publications, Inc.

Gardiner, T. (2002). *Senior mathematics challenge.* Cambridge: Cambridge University Press.

Gardner, M. (1956) *Mathematics, magic and mystery.* New York: Dover Publications, Inc.

Goldenberg, E. P., Shteingold, N. & Feurzeig, N. (2003). Mathematical habits of mind for young children. In F. K. Lester (Ed.), *Teaching mathematics through problem solving: Prekindergarten-Grade 6* (pp. 15-29). Reston, VA: The National Council of Teachers of Mathematics, Inc.

Haese, R. C. & Haese, S. H. (1982). *Competition mathematics.* Adelaide, S. A.: Haese Publications.

Indurkhya, B. (1992). *Mataphor and cognition: An interactionist approach.* Dordrecht: Kluwer Academic Publishers.

Levine, M. (1993). *Effective problem solving (2nd Ed.).* New Jersey: Prentice Hall.

Lipson, J. I., Faletti, J. & Martinez, M. E. (1990). Advances in computer-based mathematics assessment. In G. Kulm (Ed.), *Assessing higher order thinking in mathematics (pp. 121-134).* Washington, DC: American Association for the Advancement of Science.

McGreevy, N. (2006). *Mr McGreevy's absolute howlers.* NSW, Australia: Allen & Unwin.

Odell, S. (1987). *Puzzles for super brains.* New Delhi: Sudha Publications Pvt. Ltd.

Posamentier, A. S. & Lehmann, I. (2007). *The (fabulous) Fibonacci numbers.* New York: Prometheus Books.

Peterson, I. (2001). Circle game. *Science news,* 159, 254-255.

Rigby, J. F. (1997). Traditional Japanese geometry. *Mathematical Medley,* September 1997, 41-45.

Salkind, C. T. (1961). *The contest problem book I: Annual high school contests 1950-1960.* Washington, D.C.: Mathematical Association Of America.

Salkind, C. T. (1966). *The contest problem book II: Annual high school contests 1961-1965.* Washington, D.C.: Mathematical Association Of America.

Seckel, A. (2000). *The art of optical illusions.* Carlton Books.

Snape, C. & Scott, H. (1995). *How many?* Cambridge: Cambridge University Press.

Stewart, I. (2008). *Professor Stewart's cabinet of mathematical curiosities.* London: Profile Books Ltd.

Tien, L. C. (2001). Constant-sum figures. *The Mathematical Intelligencer,* 23(2), 15-16.

Vakil, R. (1996). *A mathematical mosaic: Patterns & problem solving.* Burlington, Ontario: Brendan Kelly Publishing Inc.

Von Savant, M. (1994). *More Marilyn.* New York: St. Martin's Press.

Walser, H. (2001). *The golden section.* The Mathematical Association of America.

Winicki-Landman, G. (1999). Stars as a source of surprise. *Mathematics in School,* 28(1), 22-27.

Yan, K. C. (2008). *The beauty and power of visualisation.* Talk delivered to Cluster XX teachers on 20th March, 2008 at Northland Secondary School, organised by Panpac Education.

Yan, K. C. (2011). *Mathematical quickies & trickies.* Singapore: MathPlus Publishing.

Yan, K. C. (2007). *More illusions in mathematics. YG (Young Generation)*, 318, 10-11.

About the Author

Kow-Cheong Yan is the author of Singapore's best-selling *Mathematical Quickies & Trickies* series and the co-author of the MOE-approved *Additional Maths 360*. Besides coaching mathletes and conducting recreational math courses for students, teachers, and parents, he edits, ghostwrites, and consults for *MathPlus Consultancy.*

Kow-Cheong writes about the good, the bad, and the not-so-ugly of Singapore's math education and of the local educational publishing industry. Read his two math blogs at **www.singaporemathplus.com** and **www.singaporemathplus.net**, or follow him on Twitter @MathPlus, @Zero_Math and @SakamotoMath.

For consulting, editorial and training services, contact him via his e-mail coordinates at **kcyan@singaporemathplus.com** and **kcyan.mathplus@gmail.com**.

Visit Kow Cheong's Facebook Pages and Pinterest Boards at these virtual addresses:

fb.com/SingaporeMathPlus
fb.com/AddMaths360
fb.com/Christmaths
pinterest.com/MathPlus

For a sub-list of his published Singapore math titles, visit
amazon.com/author/yankowcheong

www.ingramcontent.com/pod-product-compliance
Lightning Source LLC
Chambersburg PA
CBHW051410200326
41520CB00023B/7180